DATE DUE			

MAN, THE PROMISING PRIMATE

MAN, THE PROMISING PRIMATE

THE CONDITIONS OF HUMAN EVOLUTION

Peter J. Wilson

New Haven and London
Yale University Press

Published with assistance from the foundation
established in memory of Philip Hamilton McMillan
of the Class of 1894, Yale College.

Designed by Sally Harris
and set in VIP Garamond type.
Printed in the United States of America by
Vail-Ballou Press, Binghamton, N.Y.

Library of Congress Cataloging in Publication Data

Wilson, Peter J
    Man, The Promising Primate.

    Bibliography: p.
    Includes index.
    1. Human evolution.   I. Title.
GN281.4.W56   573.2   80-12724
ISBN 0-300-02514-9

10   9   8   7   6   5   4   3   2   1

To Robin and Ian

It seemed worthwhile to try how far the principle of evolution would throw light on some of the more complex problems in the natural history of man. False facts are highly injurious to the progress of science, for they often endure long; but false views, if supported by some evidence, do little harm, for every one takes a salutary pleasure in proving their falseness; and when this is done, one path towards error is closed and the road to truth is often at the same time opened.

Charles Darwin, *The Descent of Man*

CONTENTS

Preface   xi

Introduction   1

1 The Primitive Primate   11

   What Conditions Make Human Evolution Possible?   11
   Phylogenetic Aspects of Hominid Brain and Thought   14
   The Significance of Upright Posture   20
   The Environmental Background of Generalization   27
   Ontogeny and Generalization   33
   The Logic of the Evidence and Its Conclusions   39

2 The Promising Primate: Structure   45

   Brainchild: The Meaning of a Complex Brain   45
   Social Organization among the Hominoids   49
   Relativity: How Are Human Relations Possible?   54
   Fatherhood and Kinship   64
   The Incest Taboo   69
   Marriage   79

3 The Promising Primate: Content   83

   Consciousness   83
   Individuation   87
   The Promise   97
   Technique: The Production of Reproduction   103

4   Dietary Considerations   111
    Thinking about Objects   111
    Hominoid Diet   115
    Evolution of the Human Diet   119
    Early Hominids: A Digest   127
    Food and the Brain   133
    Men of Good Taste   140

Conclusion   148

Appendix: The Metaphor of Kinship   157

References   171

Index   179

PREFACE

In the past few years, the findings of biology, and particularly genetics and ethology, have led to a profusion of books and articles seeking to locate the explanation of all human activity in the operation of biologically determined processes. Human culture, it is argued, is but the outcome of genetic programs, or "biogrammars," and this being the case, we can now engage in the objective study of human activity, approaching man's behavior as we would that of atoms, chemicals, magnetic fields, animals, or plants. We can study human beings as if we were not human beings. This is what Hobbes tries to do by applying the "resolutive-compositive" method of Galileo and Harvey to the study of human society. It is what Hume thought to do by taking Newton's method of inquiry, experimental reasoning, and introducing it to moral subjects. The result was an attempt to demonstrate that everything human about human beings follows from the nature of their passions, emotions, and feelings and from the impressions gained from their sensory apparatus. In other words, that which is human can be reduced to that which has been naturally bestowed. Modern theories along "scientific" lines represent, in broad terms, a restatement of the argument, replacing what were earlier thought to be irreducibles (passions and sensations) with what are now thought to be irreducibles (genes and genetic programs).

Now, this point of view is not wrong—no point of view can be wrong—and it is not my intention to dismiss it. It is partial, however—correct only insofar as it informs an inquiry whose purpose is to look at human beings as if people, animals, plants, and machines could be considered things of the same sort. This approach is suited to such human practices as medicine, nutrition, or brainwashing. But it does not apply to "subjective" issues—such as why human beings think it important to try to cure each other, feed each other properly, or brainwash each other. In the present case it does not apply to why we think about evolution. Reductionism is viable only as long as we remain outside the subject of study, but as soon as the fact that we study a subject becomes a critical part of the enterprise, reductionism is neither an independent nor a sufficient answer. Organisms are not reduced to genes, it is *we* who reduce organisms to genes and part of what we want to explain is why this is so. The short answer to this problem is that such thinking constitutes human evolution. The following essay is an elaboration of this argument.

No reader, I hope, will think that the present book seeks to describe and explain the origins of the human species and its cultures. For this would suppose a factual essay. I do, of course, make use of some facts, but it is at no time my intention to summarize or synthesize the known facts of human evolution. The facts I invoke serve only to guide and provide faint boundaries for a purely speculative argument. It speculates not about the origins of human society and culture but about the conditions that made such origins and their development possible. My intellectual debts are to seventeenth- and eighteenth-century philosophers and in some sense to Marx. They provided my starting point.

It remains now for me to acknowledge more personal debts. An early but vital draft was written while I was on a six-month sabbatical leave from the University of Otago, part of which I spent as a visiting fellow at Churchill College, Cambridge University. My thanks to these institutions and their decision-making members, who made this leave both possible and enjoyable. I am particularly grateful to Dr. Ray Abrahams of Churchill College for his efforts on my behalf.

Versions of several parts were published in *Man,* and a version of another part was rejected. Both acceptance and rejection helped enormously to encourage and clarify my thinking. The original germ of this book, which resulted first in a paper entitled "The Promising Primate" (Wilson 1975), was planted during seminars at the Australian National University. I much appreciated my stay there and especially the stimulus of debates with Prof. Derek Freeman. Whatever clarity and precision have been achieved in this essay are due to the enthusiasm and editorial skills of Ellen Graham and Marcia Brubeck, of Yale University Press. Elizabeth Moore typed the numerous versions of manuscript. And finally I have to thank Joan, Marc, and Duncan for letting Dad get on with his writing. This book was written during the first two years of life of our twin sons, Robin and Ian, to whom it is dedicated—with all my love.

INTRODUCTION

In the following pages I will consider human evolution from the point of view of social anthropology. My aim is to draw attention to the importance of nonmaterial, social factors in human evolution and development. One can immediately object that social factors leave behind no evidence that can serve as a concrete subject of study, and with this view I agree. There is no possibility of discovering any reliable data concerning the original state of human social organizations. Nevertheless, they have a structure; they are systematic arrangements of ideas treated as social facts, and a system follows certain rules and principles that are logically, if not temporally, basic. It should therefore be possible to discover the bases for human social culture and then to associate them with the earliest members of the genus *Homo*.

From an evolutionary point of view, humans are one among a number of species composing *Homo*, which in turn must ultimately be recognized as a constituent of the order Primates. Since we know that *Homo* is an evolved member of the order and that it emerged comparatively recently, we are in a position to assert that human culture emerged and began its development at some point in time; that is, at one period there was no human species, hence no culture, while at a later point there was. We may also state that within the order Primates, this one genus has developed culture, whereas all other members of the order have not.

These assertions provide us with a context in which to examine

1

human evolution and to frame pertinent questions. The central question is, What conditions made it possible for one primate genus to develop culture at a certain period of time, namely the Plio-Pleistocene era, in what is now Africa? This question seems to me the most important one not yet asked, let alone answered, in the study of human evolution. Some amplification of this rather presumptuous statement may be needed, especially in the light of the current influence of sociobiological theories and the controversy surrounding them.

In its most extreme version, sociobiological theory claims that all human activity, including the cultural, is ultimately determined in its general form by genetic programs. No sociobiologist would wish to assert that every definable human trait has a corresponding gene or set of genes determining it. Rather, within the limits of species genotypes, genetic programs exist that may be said to contain the potential for specifics of human culture (E. O. Wilson 1978:47). Human culture is, therefore, immanent. I myself can see no real objection to this view, and I am prepared to acknowledge that at least in its general forms, all aspects of human culture must in the long run be traceable to the limits and potentialities of the genotype.

To say, then, that the origins of our culture lie in the genes is broadly true. The earliest hominid to make tools could not have done so without the genes that resulted in the opposable thumb and the enlarged complex brain. But we must ask whether any other conditions were necessary before tools could be made. The human genotype provides a potential; still, evolution itself occurs only if and when the potential is transformed into the actual. Thus the human genotype belongs to the realm of possibility, together with other conditions that enable the import of the genes to be realized. No matter how demonstrable it is that human traits are genetically determined, such a demonstration will not provide the answer to the question "How was the evolution of the genetic potential made possible?" Biological theory does not and cannot answer questions about the *evolution* of human culture because the Darwinian notion, and probably the mechanics, include biology as only one among several factors.

The biology of a species is part of its life, which is related to context, or the changing structure of the environment. A species survives by perpetuating itself in the world at large. In neo-Darwinian theory, the gene provides the potential, which the conditions of the environment select or reject by the simple and ultimate strategy of allowing or preventing survival. If a trait develops accidentally, by genetic mutation, the organism manifesting it will survive if the mutation proves adaptive. Accident may mean that the creature takes the initiative in developing the trait, but such initiative can be sustained only in relation to the context. On the other hand, the surroundings may change and provide a climate that selects for a genetically determined feature that might otherwise have remained dormant and useless. Here the environment takes the initiative, but the overall conditions remain the same. Evolution in its broadest terms is made possible by the interrelation between the conditions of the setting and the conditions of the organism.

In fact the evolution of a particular species is a fiction, for evolution is but the ongoing of life, a seamless web. Every species is part of the environment, hence all species *are* environment. It is therefore only if we make an incision in the web that we can study evolution analytically. In this essay I shall extract the genus *Homo* and hold it as the center of attention in order to speak of the relation between the genus and the environment and hence to formulate the problem for investigation. I shall assume that the conditions making human evolution possible include the existence of genetic programs that hold the potential for forms of activity; I prefer here to focus on the necessary environmental conditions. I shall propose that since evolution as a totality is ongoing, the environmental context of any given species may be considered problematic, that is, as requiring adjustment at some time or other. This problematic is not a singular entity but rather a complex and simultaneous totality. Still, any investigation of it must of necessity analyze, separating out the parts and treating them "as if" they were separate and "as if" they demanded separate solutions. Thus although I consider questions of kinship and social organization (that is, reproduction) before I consider those of food and diet (production), the two aspects belong together as part of a complex whole, and they confront the species as simultaneous problems. Hence their solutions are also interrelated and interdependent.

This approach to human nature and its evolution has its roots in the philosophical anthropology of Hobbes, Locke, Rousseau, Hume, and especially Kant. For all these thinkers, the central question has been, How was a human species possible? And each has argued by starting from conditions of contradiction.

Hobbes contended that all human beings are born as individuals, each of whom has the natural right to secure the means of his survival in the best way possible, that is, by improving his own welfare, hence serving his self-interest. As is well known, this course of action gives rise to an impossible, contradictory situation—the war of all against all, since if every individual is bent solely on his own prosperity, all men will be in competition and can never rest secure with their individual gains. But the human race manifestly *has* survived and has even improved its condition, a feat that could have been possible only if men found a way to overcome their natural condition by ensuring their own welfare through the well-being of others. This end was achieved through a contract—or rather, the naturally contradictory state of man is resolved by replacing it with the apparent paradox of the social contract which best serves self-interest by serving the interests of others.

It was Rousseau who exposed the major fallacy in Hobbes's argument, the assumption that before the existence of society, men were social and moral, whereas natural man could be neither moral nor immoral, but simply asocial and amoral. If the original condition of humanity was individual self-subsistence, then what foundation is there for proposing that any individual will have an interest, positive or negative, in any other individual? For Rousseau, human evolution is made possible in part by the conditions of human nature and in part by the ecological conditions in which the species is placed.

The natural condition of the human species is its eclecticism, its generalization:

> Whereas every species of brutes was confined to one particular instinct, man, who perhaps has not one peculiar to himself, would appropriate them all, and live upon most of these different foods which other animals shared amongst themselves: and thus would find his subsistence much more easily than any of the rest. [Rousseau 1754/1973:48]

But if life was so easy for natural man, how are we to explain his dissatisfaction, evident from the species' efforts to improve its situation and enhance its chances of survival? Rousseau proposes that as man's needs, tastes, and preferences became specific and well defined, the obstacles and barriers to their satisfaction became more difficult to surmount:

> I suppose man to have reached the point at which obstacles in the way of their preservation in the state of nature show their power of resistance to be greater than the resources at the disposal of each individual for his maintenance in that state. That primitive condition can then subsist no longer; and the human race would perish unless it changed the manner of its existence. [Rousseau 1762/1973:173]

But as our needs expand, so do our desires—and so too do the differences between men. The expansion produces a peculiarly human dilemma, phrased as a contradiction:

> Our needs bring us together at the same time as our passions divide us, and the more we become enemies of our fellowmen, the less we can do without them. Such are the first bonds of general society. [Rousseau 1762/1973:155]

Rousseau's theory accepts the essentially contradictory nature of the original human condition, for which Hobbes had argued, but shifts it forward, as it were, from a social man to a man whose nature bears traces of sociability. And he welds the new concept to the ecologically contradictory circumstances to which Locke had first drawn attention.

Like Hobbes, Locke affirmed that all men are born with a title to perfect freedom and uncontrolled enjoyment of all the rights and privileges of the law of nature, equality with any other man. In particular, this means that the "Earth, and all that is therein, is given to Men for the support and comfort of their being" (Locke 1690/1960:304). How, then, is it possible for some men to acquire rights and ownership of land and its resources by competing against others to ensure their own individual survival?

Locke's answer to this paradox is that possession or property begins with that which "everyman has . . . in his own person. This nobody has a right to but himself. The labour of his body and the work of his hands, we may say, are properly his." When a man mixes his labor with material that nature provides, he contributes something that is his own, and the result thereby becomes his property.

The advent of property, however, leads to a possible Hobbesian state, for human nature is essentially violent:

> The pravity of mankind being such, that they had rather injuriously prey on the fruits of other men's labours, than take pains to provide for themselves; the necessity of preserving men in the possession of what honest industry has already acquired, and also of preserving their liberty and strength, whereby they may acquire what they further want, obliges men to enter into society with one another. [Locke 1689/1824:42]

Thus men must give up their predatory rights, that is, they must observe the interests of others in order best to secure their own interests. In this way, the uncertainty brought about by the contradictory nature of the human situation can be relieved; self-interest in this new guise makes man

> willing to quit a condition, which however free, is full of fears and continual dangers: and it is not without reason, that he seeks out, and is willing to join in society with others who are already united, or have a mind to unite for the mutual preservation of their lives, liberties and estates, which I call by the general name property. [Locke 1690/1960:368]

Hume takes over the arguments based on contradiction that were put forward by Hobbes, Locke, and Rousseau, but with his unerring sense for the exquisite, he refines the contradictions.

Hume's first argument from contradiction is that of Rousseau—that man is generalized and therefore wants more than other animals but is less able to procure his satisfaction with the natural means at his disposal:

> Of all the animals, with which this globe is peopled, there is none towards whom nature seems, at first sight, to have exercised more cruelty than towards man, in the numberless wants and necessities with which she has loaded him, and in the slender means, which she affords to the relieving these necessities. . . . In man alone, this unnatural conjunction of infirmity, and of necessity, may be observed in its greatest perfection. [Hume 1739/1888:484–85]

The resolution of this conflict is made possible by society—in fact *is* society. But society becomes both possible and necessary as it resolves the Lockean contradiction of property, for only property, of all that humans may be possessed of, can be taken away by others. If there were enough resources in the world to go round, there would be no need to take other people's property away, nor any desire to do so. Thus, contra Locke and anticipating Malthus, Hume points to the scarcity of natural resources that may be transformed into property, which may be expos'd to the violence of others, and may be transferred without suffering any loss or alteration; while at the same time, there is not a sufficient quantity of them to supply every one's desires and necessities. As the improvement, therefore, of these goods is the chief advantage of society, so the *instability* of their possession, along with their *scarcity,* is the chief impediment. [Hume 1739/1888:487–88]

The overall formulation of the paradox that underlies the Humean platform supporting society is a synthesis of Hobbes, Rousseau, and Locke. For Hume suggests that through property, and in the service of justice, society emerges as that association of people *together* by way of an agreement to keep their properties, both in themselves and in goods, *separate.*

All these forms of the contradiction that makes society possible and makes the species human are summarized in Kant's dictum that the character of the species is one of "unsocial sociability," since

taken collectively (the human race as one whole) it is a multitude of persons, existing successively and side by side, who cannot do without associating peacefully and yet cannot avoid constantly offending one another. [Kant 1784/ 1974:91]

Further, this statement has its roots in a contradiction identified by Rousseau and based on his view of a somewhat capricious nature that "seems to have taken pleasure in the strictest economy and to have measured out the basic animal equipment so sparingly as to be just enough for the most pressing needs of the beginnings of existence" (Kant 1784/1970:43).

It can be seen, I think, that theories about the nature and origin of human society and about the attainment of human status by what is otherwise a mere animal species are rooted in the formulation of contradiction. As Hobbes readily observed with respect to his argument, it is doubtful whether the human species was ever engaged in a literal war of all against all, but as far as he could see, expressing the matter as a contradiction is necessary for any explanation of the beginnings of human society, or more accurately, for any representation of those beginnings. This way of reconstructing "origins" describes not a historical fact but a present one, namely, that at every point of analysis we are confronted by uncertainty, that this uncertainty breeds fear and anxiety, and that in order to cope with fear, anxiety, and uncertainty we have to try to understand and resolve. To state the matter a little differently, the Hobbesian war of all against all is not so much a state that ever existed as a historical fact as it is a state that exists as an ahistorical fact. For the contract, exchange and society; law, government, and justice; and the striving to attain the categorical imperative are simply the ways and means of repressing violence and self-destruction, but not of removing them.

The critical stance adopted in this essay is analogous to that taken by Marx toward history. Human activities and their results must be considered in relation to their material conditions. I propose to do this for prehistoric conditions, or even more specifically, for the time preceding human emergence. Beyond adopting this position I make no claim to alliance with Marx or with anyone else. Although I am, in a sense, reviving the speculative ideas of seventeenth- and eighteenth-century philosophers about the foundations of human emergence and existence, more than two hundred years have passed since their day. In that time a great deal of material evidence has been discovered about the presence, appearance, and situation of the first human beings and their immediate predecessors. Much has also been learned (and is still being learned) about the nature of those species most closely related to human beings, and the advance of knowledge has made the definition of exclusively human characteristics more difficult and complex. Above all, the Darwinian theory of evolution has been propounded and modified and has established itself as the foundation of modern thinking about the nature and organization of life. Wherever these advances bear on the speculative ideas of philosophical anthropology, they must be taken into account. When the facts do not accord with the speculation, then it is clear which has to go, but we can retain the ways of regarding and interpreting facts, the means of relating them such that their apparent independence and isolation are counterbalanced by our sense of how they belong together within a process.

1 THE PRIMITIVE PRIMATE

What Conditions Make Human Evolution Possible?

Once the species *Homo sapiens* emerges, human evolution becomes far more a story of the development of "culture"—an inventory of products made possible by brainpower and the aptitude of the body for putting thought into action. Human culture may be said to comprise (1) a hard, material component; (2) a soft, institutional component; and (3) an ideational component that unites the first two. The material part tends to leave behind traces such that future generations can, to some extent, reconstruct large chunks of the population's way of life and its attendant ideas. But the soft part of culture—kinship, political organization, religion, law, and economic organization—disappears with its practitioners unless and until they develop some material means of describing or representing their activities and so leaving a record.

Most of our knowledge about prehistoric human societies concerns their material culture and the inferences we are able to make about the technique or engineering that produced the artifacts. In addition, we are able to consider the hard remains of prehistoric human individuals, their skeletons, and make comparable inferences about their anatomical engineering capabilities. From the pelvis, leg and foot bones, and skulls we can infer upright posture; from the head we can infer brain size and even some external convolution patterns of the brain; from teeth

we can infer dietary preferences. But as to the thoughts of prehistoric people and the ways in which they organized their activities,
bones and artifacts tell us little. Yet judging from our own orders
of priority and from those of numerous contemporary human
populations, the soft component of culture is at least as distinctive
as the hard part in creating and defining what is human. So any
description of human evolution based solely on material remains
will not be very satisfactory. On the other hand, the legions of just
so stories put forth to account for the origins of known patterns of
human activity, while doubtless serving a need to satisfy human
curiosity, are not calculated in the long run to further our understanding either of the past course of events or of the present
nature of human nature.

Particularly unsatisfying are the "suddenly must have" theories,
in which a primate species suddenly must have stood upright or
suddenly must have invented language or suddenly must have
discovered hunting and cooperation. There is no reason to think
that human evolution has departed so radically from that of other
species as to have been the subject of miracles or that human
evolution has not been subject to countless interlocked developments and changes, rather than privileged with one or two
momentous alterations. Nor is there any reason to think, as these
theories imply, that the initiative for human development has
always lain with the species. Evolution in general is a process of
natural selection, of morphological adaptations to contextual
conditions that facilitate the successful reproduction of organisms. The environment, using that term for the moment in its
broadest sense, has been as instrumental as the species in further-
ing evolution.

The ever-changing ecology presents a problem to whichever
species, hominid or otherwise, we choose to isolate for consideration. If an animal succeeds in adapting to a particular environment, then from the point of view of any other species occurring
there, such an adaptation, as a change, is the essence of the problem. So if we are concentrating on hominid evolution, we need
also to be aware that the setting in general and in its particulars

demands of the species an adjustment, which in turn constitutes its adaptation.

Now, it is true that most problems allow for more than one solution. Faced with a barrier of water, an organism can swim across, float over on a log, build a bridge, jump, perhaps find a convenient vine for the purpose of swinging across, or look for a stretch shallow enough for wading. An obstacle imposes conditions, however, and to be met they must be reconciled with the limitations and capacities of the organism. If the anatomy does not permit swimming or flying, then those two possible solutions are out. The restrictions of the environment, taken in conjunction with the limits of the creature in question, suggest a "framework of possibility." Since we can be wise after the event with respect to human evolution, and since we know that the hominid species developed powers of rational thought, made efficient artifacts, invented language as speech, subsisted by hunting and gathering, and organized individuals, we can ask: What sort of environmental and biological characteristics of the species made these solutions possible? To answer this question by no means ensures that such solutions were in fact necessary, for there may well have been options that the organism simply did not happen upon or that, for all we know, were rejected. But at the same time, the possibilities may have been such that the action taken was necessary, and we might be able to show as much. My question, then, is: What conditions, particularly in the late Pliocene, early Pleistocene era, permitted human evolution?

At the outset I will also note that although we are accustomed to separate material culture and its artifacts from the software of "social culture," there is little or no justification for this practice. A prehistorian may by rights be concerned with stone tools and wish to infer from these the technique of their manufacture. A social anthropologist will by rights be concerned with the system of kinship in a community, so that he may understand the principles of its social organization. But with respect to human evolution, that which makes possible the manufacture of stone tools may also make possible the development of a kinship system. I mean not that chipping a stone will produce a mother's brother but that the mode of thought, conceptualization, and connection necessary for the one may also be necessary for the other, and that at a broad level of abstraction, the form and power of thinking evident in stone tool technology may suggest how kinship was possible. We may say, then, that if we can determine in the environment of a hominid a problematic related to social organization, we can point out where and how kinship was a possible solution that could be arrived at within the limits of the species and the environment.

To state this a little differently, while we have tended to think of material culture as comprising artifacts and have assumed that they come into existence through manufacture, we have also tended *not* to think of social institutions as also requiring invention and creation. Yet the soft parts of culture are as man-*made* as the hard parts and must be brought into existence by some sort of technique. There is no reason to think that the manufacturing technique used to create material things, especially the cognitive or cerebral aspects of it, is so exclusive as to call for ideas radically different in nature from those applied to creating social artifacts. In short, if the human brain is shown to have been capable of planning the production of complex artifacts, directing human limbs and eyes for this purpose, then why should we think that it could not plan and direct the production of an organized social life? The two projects obviously differ in the nature of the materials to be fashioned and the ends to be achieved, which of course locate the powers of thought in different areas of the environment, but that need not deter us from trying to find the common characteristic.

Phylogenetic Aspects of Hominid Brain and Thought

There is hardly any need to argue that the powers of thought, imagination, invention, and control—the brain as mind, in other words—are the salient aspect of human evolution. Whatever the outcome of our journey, our brainpower has thus far made the difference in our struggle to survive.

Another distinctive feature of the species has been its ability to populate most of the regions of the world, that is, to adapt to an enormous range of environments. This suggests that human anatomy and physiology have attained an extraordinary degree of flexibility or generalization. Man is able to survive and reproduce under extreme climatic and ecological conditions, to cope with an enormous topographical variation, to extract and digest a virtually unlimited variety of food, and to withstand sudden changes.

In these two respects—development of the brain and a generalized morphology—the human species is characteristic of the primate order of mammals, only more so.

> One of the outstanding features of the order Primates is that its members cannot be defined by any single or peculiar character, but rather only by a combination of characters, any of which may be found in members of certain other mammalian orders. [Straus 1949:200]

Straus goes on to quote Zuckerman:

> These morphological characters are generally believed to represent a primitive mammalian condition, so that it may be truly said that the Primate, except for its general tendency to cerebral development, is a relatively non specialized mammal. [Zuckerman 1933, quoted in Straus 1949:200]

As the primates are to mammals, so is *Homo sapiens* to the primates: an even more generalized species specializing in brain development.

Yet the apparent sophistication of the human brain is so great that it seems to be more than simply the continuation of a primate tendency. One is tempted to wonder whether or not some extraordinary difference attends this development, making it a unique and extreme specialization. Before considering such a possibility, it is worth seeking an explanation for the observed but unexplained correlation between primate cerebral development and morphological generalization, for such an explanation may serve as a prelude to understanding the apparently peculiar case of *Homo sapiens*.

In the *Penguin Dictionary of Biology* (Abercrombie 1957) the term "generalized" is defined as "not specialized." I shall therefore take the liberty of simply reversing the definition for "specialized" in order to define "generalized":

> Having no special adaptations to a particular habitat or mode of life with little divergence from presumed ancestral forms; and which tend to place few restrictions on the range of habitat which can be occupied and the variety of mode of life which can be followed and hence, it is assumed, place few limits on evolutionary flexibility.

Being generalized or primitive mammals, primates can, in comparison with other mammalian orders, range freely over large areas of the world and show a variety of life-styles. They live in forests, plains, on highlands and lowlands, in hot moist and hot dry climates, in the old world and the new. Some are solitary, others are gregarious; some are nocturnal, others diurnal. But they do have their limits. Primates other than hominids are not found in areas with severely cold temperatures, whether these extremes are seasonal or permanent. Hence they are not found in northern or southern temperate and arctic or subarctic zones but are confined more or less within the tropics. No primate can be described as carnivorous, although it is important to note that some can, if need be, eat meat; nor are they aquatic. The human primate is the one species that has transcended these general limits over the course of time.

Various genera and species of primate have developed certain specializations concerned almost exclusively with locomotion. New World Cebidae travel with the aid of a prehensile tail; gibbons in particular have specialized in moving by brachiation; chimpanzees and gorillas cross the ground by knuckle walking;

baboons are monkeys that have adapted to plains living, particularly by adopting a semiupright posture for feeding and an opposable thumb and forefinger allowing them a precision grip. All of these specializations to some extent impose limits on their species, but not so much that alternate forms of locomotion are impossible. The knuckle walkers are still excellent climbers; brachiators can, if need be, move at some speed on the ground. Human specialization has consisted of upright posture and bipedal locomotion, but interestingly, these features in a sense extend a number of specialized behaviors engaged in by other primates. It is true that no other primate walks upright all the time, but the great apes in particular do so sometimes, and baboons feed and shuffle in a semiupright position. All primates use their forelimbs, rather than their mouths, to obtain plant food, as is characteristic of other orders, so that the freedom of hominids' forelimbs for tasks of extraction, manipulation, and carrying simply extends a primate trait to its logical conclusion.

By the same token, the human forelimb is capable of performing the important support and locomotor functions carried out by its counterpart in other primates. Humans can crawl, for example, and infants do so extensively; this is often important for activities such as hunting, where uprightness can be a disadvantage. Reports of "wolf children" suggest that humans can go about on all fours, and keep up with the pack. Man can climb, but not as well as monkeys and apes; man can move quite quickly in the open, but not as fast as quadrupeds. In other words, the reduction of the forelimb and the development of the hind limb do not exclude the human being from modes of locomotion carried out more expertly by other primates.

Uprightness and bipedalism, though in one sense a specialization, also represent an extreme generalization of forms of primate locomotion. The human anatomy shows characteristics some of which can be found throughout the order, but no other primate species seems to have as much in common with the others as the hominid does. In fact, if we take an even broader view, the human species seems to share more traits with different orders and phyla than any other species does. We can learn about human responses from the responses of rats and pigeons; about our chemistry from mice, pigs, and monkeys; and about our behavior from chimpanzees. If primates are the most primitive and generalized of mammals, then human beings are the most primitive and generalized of primates. This statement applies particularly with respect to the "specialized" human posture and gait.

But what do we mean when we describe the character of a species as "generalized"? And in particular, what is the significance of the association between this generalization and cerebral specialization? Why is primitiveness of primate morphology correlated with enlargement and development of the brain, which is of such great importance for human evolution? Any understanding of that evolution in its specifics must be able to account for such a phylogenetic correlation, possibly even to the extent of examining whether the one may be the cause of the other.

Referring back to the definition of "generalization," we see that it stresses function. If specialization adapts an organism to a particular habitat by formulating a specific mode of life, then generalization, in contrast, precludes specific adaptation and any fixing of a mode of life. The generalized locomotor morphology of the human primate excludes any conditioning of the species to a specific environmental niche and so leaves man free to find his means of survival in a variety of settings and in the environment in general.

Here we must pause, however. For the environment does not exist in general: there are only specific environments—mountains or valleys, forests or plains, woods or savannahs, hot and moist, hot and dry, cold and wet, cold and dry, windy and calm, and so forth. Whereas a specialized animal can simply rely on its surroundings, finding subsistence by virtue of its adaptation, a generalized organism must fashion from its adaptive potential some mode of life suited to both the ecology and the capabilities of the creature itself.

A generalized organism must therefore direct general capabilities into forms of specific activity. And if circumstances change, the strategy of adaptation must change. The weakness of generalization is also its strength: not being specially fitted to environmental niches, an individual or a population must struggle to define its potential so as to survive, taking account of all eventualities, from the lie of the land to the available forms of sustenance to the nature of the competition. Compared with highly adapted organisms, the generalized creature is at a disadvantage, but this is offset by his ability to adjust and/or seek out alternative niches. The more generalized an organism, the greater the opportunity for experiment—in fact, adaptive strategy requires trial and error, search and find. The lessons of success must be retained, and the warning of failures must be kept in mind.

Thus all animals use the brain to control their movements and activities in an environment. Even the most specialized creature directs its organs toward food or avoids predators by processing sensory information and transmitting directions to limbs and organs (Gibson 1966). But whereas in specialized organisms the brain and the body are identically specialized, this cannot be so for generalized creatures. The trials and errors of a generalized creature engage its brain in much more complex operations of storage and retrieval, coordination, perception, and selection of information about the environment. The brain, as the executive director of the body, must relate the general capacities to specific modes of operation, organizing and reorganizing to meet different contingencies. A creature that depends on four legs for locomotion on the ground also relies on four legs for climbing, and the extent of reorganization is minimal. A generalized two-legged creature must reorganize its limbs and body for climbing using four limbs. The brain is the instrument that solves these problems.

The nub of my argument, then, is that a generalized morphology in primates, when understood in terms of functional adaptation to environment, is necessarily concomitant with cerebral development, and this feature is integral to hominid phylogeny.

As the most generalized of primates, the hominids show the greatest degree of cerebral development. Generally speaking, then, the human brain cannot be considered exceptional in terms of phylogeny, representing, as it does, the end point in a logical progression.

But this argument, by its very general nature, is not too satisfactory, although it is valuable as it shows the nature of the necessary evolutionary problem (and of the adaptational strategy) facing generalized primates and hominids in particular. We still need to explain why the hominid brain developed in the particular directions it did.

The crucial truth suggested by our reasoning thus far is that hominid adaptational strategies, and hence evolution, are teleological and functional. In other words, human morphology may be considered sufficiently generalized to possess potentiality great enough to meet with most of the contingencies and variations that the environment, in its specifics, is likely to provide. It is a matter of organizing morphological structure to perform specified functions, relative to a purpose. For example, the human arm is capable of throwing, but this function of the arm has to be organized and defined from the potential. Such organization can be effected only by calling upon the capacity of the brain for coordinating sensory information—particularly vision and the measurement of direction and distance—with motor efficiency. These are two activities that the brain undertakes in any case, and among hominids who walk upright it performs them with greater precision and organization than among nonhominids.

The Significance of Upright Posture

Upright posture and bipedal gait appear to have become a specialized trait complex of hominids, distinguishing them from nonhuman primates. The question arises as to whether uprightness is correlated with the apparent specialization of the human brain. In fact, upright posture and bipedalism are not so much unique as they are a logical conclusion of a tendency increasingly apparent in higher primates, notably among the anthropoids. And in the same way, the hominid development of the brain is the outcome of a progressive and more and more apparent tendency among primates in general and anthropoids in particular. Let us consider two specific cases. Chimpanzees spend quite a lot of time on the ground, and gorillas less so. Both have been observed to walk or sit upright and to perform certain actions while standing or walking on two legs. Jane Goodall's observations and the schooling of chimpanzees in the use of sign and symbol language permit no doubt of the chimpanzee's cerebral capacity and its relatively close approach to that of humans. Upright posture, bipedalism, and brain development seem to be connected, parallel tendencies in the anthropoid primates. If we wish to explain the emerging powers of the brain for rational thought, therefore, the conditions and functions of these characteristics deserve some attention.

Mechanically speaking, the elevation of a creature from a horizontal to a vertical position has a number of consequences. As most textbooks explain, the upright creature can rely less on the sense of smell for the location of food and enemy and must depend more on sight and sound. When the head is raised several feet above the ground, the extraction of food by the teeth and jaws is less efficient and practicable, but the opportunity to use the forelimbs and hands for this purpose is enhanced. Notably, among hominids the snout area decreases, and the extracting teeth—the incisors and canines—are likewise reduced. The hands become more mobile and precise, increasing in importance as organs of touch. The most significant consequence of uprightness is the necessity for improved coordination of various parts of the body that become functionally more differentiated. The hind limbs and the forelimbs become separated in function such that the former must take the latter to the source of food. This means that vision, rather than smell, must be coordinated with locomotion—the eyes must direct the limbs, which lead the arms to food. Eyes and arms together must judge distance, depth, shape, and form to enable the forelimbs to extract. In addition, vision must team with smell and taste, so that what is first seen rather than smelled can be described to the other senses as edible, dangerous, inedible, and so forth. Whereas among lower primates the mouth and teeth are the main implements of extraction, with uprightness the hands and fingers assume this role and accordingly must develop mechanical, precision functions. Again, such a change, like all the others, demands a corresponding, if not prerequisite, development of the brain, notably in particular areas, and especially in the matter of increasing the coordination and connections between these cerebral areas.

Now, this grossly simplified description of a process can certainly be amplified with facts and specific references to the alliance of brain areas with motor and sensory activities. But this account explains nothing, for in fact the forelimb's change of function from support to carrying and manipulation *entails* some alteration or addition in the brain. The development of the forelimb for manipulating purposes is the visible part of modification of the brain for the same purposes. Any attempt to claim that the brain "develops" *because* of upright posture, or even that upright posture develops *because* of brain development, is no more than an analytic statement that makes clear the several parts in what may be seen only as a unified phenomenon. Quite apart from the Lamarckian overtones of such "explanations," they do not take us very far toward understanding what is happening in hominid evolution; they simply state what has happened, which in a sense we know anyway. An explanation, which should tell us more than we know at the beginning of an inquiry, must begin with a search for the unknown. Let us state the problem once again. What conditions of hominid evolution could have made possible upright posture, bipedalism, and cerebral development to the human level?

Perhaps the commonest explanation offered by theorists is the "coming down from the trees" hypothesis. In Miocene/Pliocene times considerable climatic changes are known to have caused forests to recede and grassy, open plains to expand. Certain groups of prehominids chose, or were forced, to forsake the forest either in whole or in part and to seek their subsistence on the plains. These customary forest dwellers were possibly gibbon-like brachiators or apelike climbers, but in the open plains there are few trees to swing or climb on, so such populations would have had to move about in another way in the open. Since the forelimbs of brachiators are longer than the hind limbs, they could possibly be used to balance a locomotion dependent essentially on the hind limbs, rather than to support the trunk, as in most quadrupeds, and this would result in a semiupright posture. On the other hand, there is no reason why brachiators or climbers could not as simply move about on all fours, that is, with both fore and hind limbs carrying out similar functions of support and movement. Since chimpanzees display specialized knuckle walking, we might suppose them descended from an ancestral form that opted for a quadrupedal stance and gait that did not rule out partial uprightness but also did not select for it. Hominids, ending with *Homo sapiens* may be descended from former brachiators that adopted a stance and gait relying on forelimbs for balance rather than for support and power.

Now, at this point we are likely to lose sight of the critical issue, which is the relation between these circumstances of climatic change and possible development of the brain. If we say that some populations of ancestral hominids moved onto the plains from the forests and gradually came to adopt an upright posture, we imply that open plains are part of the cause of uprightness and therefore that brain development followed alterations of posture. But the brain controls posture and locomotion, and unless it is so structured as to be able to operate a limb in one way rather than another, nothing is going to happen. And the same may also be said of limbs: unless they are structurally capable of alteration, no alternative posture or locomotion is possible. Limbs and brains have to operate within limits.

The central point is not that some prehominids moved on to the plains permanently or temporarily or that they lived on both forest and plain and therefore developed upright posture. What is significant about these ecological changes for hominid evolution is that a wider range of options and possibilities is presented, or to put it another way, that a familiar situation becomes, in the course of time, a problematic one. Changes that occur in the environment and affect phylogenetically generalized species are factors that enhance the already problematic aspects of their existence. Populations of species presented with widened environmental prospects are also presented with widened environmental problems. We are concerned less with the fact that the recession of the forest area and the expansion of grasslands presented alternatives and "either/or" necessities than with the truth that new ecological specifics came into being such that the prehominid general potentiality was faced with the necessity of organizing to meet those specifics. And this is a cerebral problem.

Within its given limits and those of morphology, primate adaptation must seek to meet specific environments by trial and error. It can be argued, therefore, that where certain ancestral anthropoids adopt a solution that culminates in uprightness, others, ancestral to the chimpanzee or gorilla, adopt a solution that leans more toward quadrupedalism. One solution, ultimately, goes further in enhancing the powers of the brain, but not necessarily to the extent that it can be said to be divorced from the phylogenetic characteristics of the primate order. After all, the chimpanzee is acknowledged by contemporary *Homo sapiens* to be the closest of the primates to man in intelligence—but chimpanzee intelligence and brainpower cannot be owing to the adoption of a permanent upright posture and bipedalism, based on plains living.

The development of the hominid brain is grounded both in hominid generalization and in the increasingly problematic nature of the environment occasioned by change. The significance of Miocene/Pliocene changes, and especially of the accelerated and extreme variations of the Pleistocene Ice Ages, is not that they lead to the necessity for specialized adaptation to specialized environments such as open plains *rather than* forests but that adaptation to *both* forest *and* plain becomes a possibility and a problem. Hominid evolution, particularly in the later Pliocene / early Pleistocene times marked by ice ages, pluvials, and stadials and concerning the genera *Australopithecus* and early *Homo (habilis, erectus)*, is more and more characteristically mosaic evolution, adapting to a larger and larger variety of specific environments. By the same token, chimpanzee cerebral development seems, in its natural setting, in keeping with a relatively less generalized mode of adaptation suited to a less mosaic range of environments. But when chimpanzees are called upon to meet ecological changes, as when they are brought to zoos and laboratories for study, their powers of problem solving (in our terms, intelligence) become more developed and impressive.

Upright posture with bipedalism is a mode of locomotion within the capabilities of a number of primates contemporary with and ancestral to man. Its development as a specialized mode can be understood as the exercise of an option open to brachiators in particular—but this exercise reflects the emphasis that enlargement and complication of the environmental problematic place on phylogenetic tendencies. The struggle toward an efficient uprightness not only involves the brain in coordination and reorganization, then, but also requires the brain to solve more problems to make upright posture work. It is not the advantages of upright posture that select for cerebral enlargement and development but the problems attendant upon it. I suspect that prehominid brains developed because of the problematic complex of factors including uprightness and the possibility of mosaic adaptation, compounded by environmental and climatic changes, whereas the chimpanzee confronted fewer problems in its adaptation because it remained more or less within the forest.

Uprightness and bipedalism, considered as mechanical adaptations, must be thought of as advantageous in the context of evolutionary theory. Their advantage must relate to the fact that they represent the general mode of locomotion best suited to the exploitation of a wide range of environments. To the extent that uprightness and bipedalism are made possible only through cerebral development and reorganization, then that is the measure of their contribution to the development of the powers of thought. But at best this statement can only mean that the new posture and gait aid in the development of a brain capable of controlling complex motor and sensory activity, without signifying any necessity for enhanced and intensified cerebral activity and thought. Such activity will occur only if an organism (and its species) has something to think about and if that something is amenable to change through thought.

Again, we can take the chimpanzee as a case in point. Presumably for millennia, chimpanzees have conducted themselves with aplomb in their forest environments, and no chimpanzee could ever have dreamed of conversing and thinking in American Sign Language or with plastic counters. But if he is captured or otherwise transported to an alien environment in which food, among other things, can be obtained by conversing, then the chimpanzee individual can concentrate his potential in this direction (and he has done so).

If entire populations or species are confronted by situations that are problematic in this sense, then, whether their source lies within or outside the organism, the struggle for their solution is the putting into action of the brain. I suspect that this entire complex situation was clarified for the primate order in general, and for prehominid species in particular, with the advent of *Australopithecus, Homo habilis,* and *Homo erectus.* I shall be concerned with different, complex aspects of the total environmental problematic that faced hominid species, so I will now consider certain general matters that are both part of and background for the details that follow.

The Environmental Background of Generalization

This is an essay about evolutionary strategy and, wherever possible, about tactics. In its totality it comprises an interaction among all parts, no one of which is necessarily more or less important than the other and no one of which, in the actual state of affairs, ever remained the constant to which everything else adapted. But this essay also considers *human* evolutionary strategy and tactics, so given the bias of interest, I make hominid species the center of attention, even though evolution has neither center nor periphery. I shall first outline the background for human evolution.

In the late Pliocene and early Pleistocene periods, the fossil record reveals the presence of hominid species in areas of Africa as far apart as South Africa and Ethiopia and in the later period stretching over to the Atlantic Moroccan coast. These fossils are assigned to two genera—*Australopithecus* and *Homo*. Each genus comprises a number of species. *Australopithecus* includes the species *africanus* and *robustus/boisei*, with *A. afarensis* having been recently introduced as ancestral to the other two. *Homo* includes *habilis* and *erectus*, with *sapiens* not emerging until the late Pleistocene. It is a matter of considerable debate as to whether these are two genera that evolved in parallel or whether one, *Australopithecus*, was ancestral to the other. In the one argument *A. africanus* is regarded as the first hominid, from which evolved the divergent lineage of *H. habilis*, on the one hand, and *A. robustus/boisei*, on the other. In the other argument a common ancestor, *Ramapithecus*, produces divergent lines of *Australopithecus* and *Homo*. Johanson's spectacular finds of Lucy and the Hadar family indicate that *A. afarensis* lived between three million and four million years ago, along with similar populations found by the Leakeys in Laetolil. Now, whatever the eventual status of these finds and of others that will no doubt be made in the future, the matters that concern us make the taxonomic debate marginal. The point is that the recognition of these fossils depends on how closely their features approximate those of "modern" *Homo sapiens*—specifically, whether the skull is vaulted and rounded so as to suggest a humanlike brain, a face that is flat rather than prognathous, a dental arcade that is more U- than V-shaped, teeth with reduced canines and chewing molars, a foramen magnum set forward in the skull, a pelvis broadened to take the weight of an upright body, legs and knee joints capable of bipedal locomotion, hands with flexible digits, and so forth. The more closely a fossil resembles the skeleton of a modern human being, the more "human" it is reckoned to be. And it is clear that fossil remains from the African Pliocene and early Pleistocene resemble the human skeleton far more than they resemble the anatomy of apes and monkeys, but they are far from being identical with modern *Homo sapiens*.

The cranial capacity of *A. africanus* is 441.2 cubic centimeters in a body weighing roughly forty or fifty pounds; that of *robustus/ boisei,* 519 cubic centimeters in a body weighing approximately 150 pounds. The *robustus* species is a later one, and although the brain size is absolutely larger, relative to body size it is less than that of *africanus*. This suggests that one species of *Australopithecus* specialized morphologically, growing larger and stronger, and that the increase in brain size was allometric and does not indicate any further cerebral development over *A. africanus*. It has been argued that *robustus/boisei* specialized in vegetarianism, specifically, in eating seeds and grasses. This contention is based primarily on the development of the molars in *robustus*, but recent detailed examination of the dentition of *africanus* and *robustus* by Wallace (1975) suggests no significant difference.

The cranial capacity of KNM–ER–1470, dated to about 2.8 million years ago, is approximately 775 cubic centimeters in a body possibly not much larger than that of *A. africanus*. This is larger than the mean size of the cranium of sample specimens of *Homo habilis* from Olduvai (640 cubic centimeters) and within the range of *Homo erectus* samples from Asia (750–1029 cubic centimeters, according to Tobias [1975:365–67] and Leakey and Isaac [1976:323]). Without going into any more detail about anatomical characteristics of these genera, the evidence of cranial capacity is enough to persuade us that hominids, especially the genus *Homo,* were clearly developing toward humanness. This is confirmed by the fossils related to KNM–ER–1470 and by the postcranial remains of Lucy, which indicate "an advanced, fully upright biped with a lower limb anatomy that contrasts in some respects with that of the robust australopithecines" (Leakey and Isaac 1976:324).

Arguably in the case of *Australopithecus,* but unquestionably in the case of *Homo habilis,* we find the first evidence of "tools." These stone tools have been designated as representing the Oldowan culture, and they include flakes struck from larger stones, some of which have been modified, while others were left unretouched. Three basic forms of worked stone predominate—polyhedral basing stone, choppers, and flake knives. Clark comments:

> There is nothing esoteric about their manufacture and they are all small implements with no "formality" about them. They show, however, clear evidence of a rudimentary knowledge of working stone for the production of flakes and chopping edges. [1976:22]

Oldowan tools have been found at Olduvai in East Africa and in Morocco and Tunisia. They are found associated with *Homo habilis,* the species that was efficiently upright and whose brain outweighed that of the contemporary *Australopithecus,* and they may very well indicate that *Homo*'s adaptation diverged from that of *Australopithecus,* taking a form that eventually enabled the one to survive while the other sank into oblivion. But I wish to draw attention to a completely different matter.

Oldowan tools use a variety of stone materials, are relatively small, and are informal. They are found with *Homo habilis* in a broader area than that of *Australopithecus*. *Homo habilis* was probably more efficiently upright than *Australopithecus*. Together, these facts suggest a creature rather more generalized in morphology, hence wider ranging in its adaptation, hence confronted by more complex and broader environmental problematics. Part of its response to the challenging relation between its morphology and different environments was to make tools, to begin to exploit the freedom and efficiency of the forelimb to improve that very efficiency, and to make up for deficiency in, for example, the teeth. In other words, tools made by hands are then used to extend the use of the hands in food-getting activities. The relative inefficiency of the teeth as tools is also compensated for. Tools are but extensions of the limbs and teeth, and in this sense their advent is an extension of the generalization of hominid morphology. What is interesting about tools is their reflexivity, the fact that they are fashioned by the hominid limb to extend the functions of that limb, thereby making it even more generalized. Although Oldowan tools may well have been used as tools, the process of making them by banging stones on rocks can also be an enjoyable, percussive experience in which the creature entertains himself—as Jane Goodall's chimpanzees did with empty cans. As important as the utility is the reflexivity, for it is an indication of consciousness on the one hand, and of a separation from the environment, on the other, in the sense that the environment, or parts of it, no longer need to be viewed as unalterable. The individual need no longer accept all the limitations inherent in the environment, or to state the matter differently, the individual need no longer be compelled to act solely according to the appearance of the environment, for he now finds it possible to change that appearance to suit his own purposes. A creature picking up a stone is not bound by its given condition—say, by the natural sharpness or shape—but now finds it possible to alter the appearance and contours. But to do so, any individual must be aware of the objective possibility that the stone can be altered and must realize the subjective possibility that he, the individual, has the capacity and the know-how to effect the alteration. Again, tools indicate that knowing one's ability to change the environment is accompanied by the idea of modification in a certain manner; that is, a tool reveals a technique that reveals an idea of what the technique will produce, which in turn indicates a knowledge by the individual that he is capable of carrying out the technique and realizing the idea. The possibility of failure or of producing an ineffective tool increases this sense of reflexivity and of separation from the environment. The raw materials of nature that may be transformed can, through the failure of transformation, come to stand not only as objects but as things in some sort of opposition to the toolmaker. These considerations will occupy us later.

Between *Homo habilis* and *Homo erectus* there is a sudden break. Between the Oldowan culture and the succeeding Acheulean culture there is a sudden break. Whereas *Homo habilis* was apparently confined to the African continent, *Homo erectus* has now been found in Africa and has for a long time been known to have existed in parts of Eurasia. *Homo erectus,* then, breaks from the bounds of Africa and from the bounds of the tropics into the rest of the world and into the temperate zones. It is *Homo erectus* that finally confronts the full diversity of the environment, hence the full range of problems presented to a species that has to marshal its general potentialities into specific forms to meet the myriad variations.

Homo erectus in Africa and Asia has an average cranial capacity of 931.5 cubic centimeters, with a range of 727–1,225 cubic centimeters (Tobias 1975). As the species name implies, there is no question that it is fully upright and bipedal. It is also associated with the Acheulean tool industry, "the most widespread and, apart from the Oldowan, the longest lived cultural tradition that we know ... The Acheulean Industrial Complex probably lasted ... for as long as 1.5 million years" (Clark 1976:29).

The Acheulean culture includes much larger tools than the Oldowan—to strike flakes from large boulders required a special technique, and inasmuch as these large flakes recur, they are evidence of a degree of formality, of standardization of a technique, that in turn implies some mode of dissemination within and between generations. By the time of the later Acheulean, in the Middle Pleistocene, there is a "greater diversity of retouched tools and an appreciable refinement in the technique employed for fashioning them" (ibid., p. 37). Wood or bone fashioning tools are used rather than stone, which results "in the removal of thinner and longer flakes and a considerably more refined end product with straight, regular sides on which much more labor and skill have been lavished than was strictly necessary to make a usable tool" (ibid.).

While Clark notes that the Acheulean culture may offer the first evidence of an aesthetic sense, it is also important to recognize that this may also constitute an unambiguous indication of the need for instruction by reference to rules and possibly symbols. Clark himself is reported (in Washburn and Moore 1974) to have conducted an experiment in which graduate students were able to make simple tools like the Oldowan after watching a demonstration but were unable to make Levalloisian tools (tools similar to those of the refined Acheulean) following similar observation. To make such tools requires maker's knowledge and not just observer's knowledge, and this entails some form of analytical instruction.

The fossilized remains of the "hard" parts of *Homo habilis* and *H. erectus* and their tool cultures present enough evidence to satisfy us that something of the nature of culture had begun—at least, material culture. These species show a progressive expansion, first, of *Homo habilis* throughout Africa, then of *Homo erectus*, whose remains are found in Africa and Eurasia. Such an expansion results from, and therefore provides evidence of, the expanding adaptability of the genus *Homo*, and this in turn may be interpreted as evidence and development of the tendency toward generalization. It is therefore more than likely that in *H. habilis*, and particularly in *H. erectus*, the strategy of human evolution itself evolved. These species, then, may be considered the subject of the discussion.

Ontogeny and Generalization

Stephen Jay Gould in 1977 wrote a fascinating account of the rise and fall of theories revolving around Haeckel's aphorism that ontogeny recapitulates phylogeny. The latter part of his book resurrects the problem of the relation between ontogeny and phylogeny and presents a stimulating synthesis of current research organized by theories of his own. The nature of human ontogeny is of singular relevance to the argument I am presenting, and Gould's work, having resolved problems inherent in earlier thinking, allows us to make sense of the ontogenetic process in human evolution in particular, although Gould's focus is on evolution in general.

I have contended that among mammals, the primate order may be considered the more generalized morphologically, psychologically, and sociologically. In one sense this means that "at birth" the species is exceedingly homogeneous and its individual members undifferentiated. But from birth on, the species and its individuals must find ways and means of adapting to the specific circumstances of the world at large, and as they do so, they will become differentiated. If the human species is the most generalized of all the primates, the most generalized order among the mammals, then this generalization of morphology, sociology, and psychology must appear among individuals. But its appearance among individuals must be described in terms of their life histories, which are ultimately accountable to genetic processes as the mechanism by which ontogeny is constructed. It must be borne in mind, however, that genetics occurs not in a vacuum but in a context of environments characterized by a process of change or evolution.

Both phylogeny and ontogeny "appear," the former, in the fossil remains of ancestors, in the circumstances of the environment in which they lived, and in the remains of the methods by which they survived in such environments (their tools). It appears when we compare the form of fossil skeletons and later anatomies or human and nonhuman morphologies. From such comparisons it seems that modern *Homo sapiens* resembles more the infant form of *Australopithecus* than the adult, more the infant form of pongids than the adult. The modern skull of *Homo sapiens* is rounded and domed, and the face flat—like the Taung's infant (an australopithecine) and the infant chimpanzee. The adult australopithecine, and even *Homo habilis* and *erectus,* show a flatter cranium and a more prognathous face, as does the adult chimpanzee or gorilla. These representative facts suggest that modern *Homo sapiens* is neotenous with respect to its hominid ancestors and its primate ancestry.

Ontogeny appears with the individual. In the earliest stages of the embryo, the human and the nonhuman primate are identical, but with growth the differences begin to appear. At birth and in the earliest years, a chimpanzee infant and a human infant are startlingly similar in both head and body. But a chimpanzee reaches childhood in two or three years, and adolescence in six to eight years, and by then the differences are marked—the prognathous face accommodating the specialized dentition, the heavy ridges to support the muscles that hold up the head, and the flattened cranium of the chimpanzee—not to mention extensive body hair and specialized limb structure—all stand in clear contrast to the still infantile human being.

Gould makes the eminently sensible, seemingly simple, but radical proposition that we measure the paedomorphism of the human individual not by comparing it with the chimpanzee or any other ape but by comparing the form of the human adult with the form of the human embryo, and that we contrast our findings with the results of a similar comparison for the chimpanzee. By so doing we render irrelevant the earlier arguments that had to posit a human descent from an ape and we show that relative to the apes, the human adult departs far less and more slowly from its embryonic form than does the chimpanzee.

In this sense the human individual is childlike, and since the earlier stages of development among all primates, even mammals, are relatively undifferentiated and homogeneous, we can add that the human species in its ontogenetic development refers to, rather than recapitulates, phylogeny. The human primate is a primitive primate.

From an evolutionary point of view, the significance of human neoteny—and this is Gould's major thesis—is that it does not require us to posit a human evolution resulting from a startling genetic mutation, a macromutation or a hopeful monster (Goldschmidt 1940). Gould's thesis is that human evolution has been marked by a retardation in the rate of growth that may be owing to the effect of changes in regulatory genes that might have been relatively minor in terms of genetic chemistry. Now, on the surface this argument would seem to be an argument for a macromutation—momentous effects emerging from minor changes. But in the human case the momentousness was not evident in the beginning, and the effects themselves have evolved over time as culture. We have no reason to suppose that *Homo erectus* was so much superior to other primates in the Pleistocene era; but from our new position, we can see that certain possibilities eventuated from the circumstances of hominid evolution characterized by a retardation of growth rates, particularly that of the brain.

That human ontogeny is marked by a retardation of growth seems beyond question, and that this would appear to be the key contributor to human evolution seems more likely than the idea that the human primate is marked, somatically and genetically, by a radical difference in kind. It has now been established that the genetic difference between humans and chimpanzees is less that one percent and falls well within the range of sibling species (Patterson 1978:173). It is hardly necessary even to establish the similarity so microscopically, since we have become so familiar with the humanlike achievements of chimpanzees, which can use tools and communicate with symbols. But the divergence in growth rate between the human and nonhuman is well marked. The human being matures sexually much later than any other primate. The age of sexual maturity for chimpanzees is nine years, for a gorilla, six to seven years, and for the human being, thirteen years (that is, for females). Total growth ceases at about eleven years for the gorilla and chimpanzee and at twenty years for the human. Gestation for the orangutan is thirty-nine weeks, for the gorilla, thirty-seven weeks, and for the chimpanzee, thirty-four weeks. For the human being it is forty weeks. Here the difference seems insignificant, but the degree of helplessness of a human infant at birth is enormously greater than that of any ape, and it is generally accepted that the human being undergoes extrauterine gestation for a period of almost twenty-one months, or eighty-four weeks—an enormous difference (Gould 1977: 368–69).

At this stage in our evolution, we have come to see that the development of the human brain has been a key factor, and here too the significant difference between the human and the nonhuman is that the human brain continues to grow and develop at a fetal rate for several years after birth. The chimpanzee cranial capacity reaches 40.5 percent of final capacity at birth, while the human cranial capacity is 23 percent of final capacity at birth. Chimps and gorillas reach 70 percent of final capacity after one year, whereas it takes the human individual three years to reach this point (Gould 1977:371–72). For this reason the chimpanzee infant's head seems very human (round and domed), but also for this reason the human head remains round and the chimpanzee's head flattens. Man has a brain that continues to grow and hence must be retained within a structure that will form with it, whereas the chimpanzee brain stops growing after a few years. While the chimpanzee brain is still growing, it is receptive and adaptive to challenges in its environment, hence young chimpanzees are able to learn to solve symbolic puzzles and communicate in a humanly devised manner or are able to figure out ways of outwitting humans to get bananas (see chapter 4). But chimpanzees at the age of about eight years become less like human beings, and their human mentors must dismiss them from experiments.

Neoteny through retardation of growth rate set in a framework of mosaic evolution and environment seems to be the most accurate summary of human ontogeny. But why retardation and why neoteny? As Gould remarks, "The correlation of maturation with loss of plasticity (mental as well as physical) has long been recognized" (1977:401). This means that growth and development are correlated with plasticity, mental as well as physical, or using the terms of the present argument, that the retarded, hence continuing, growth of the human individual as species member and its corresponding sustained "youth" is consonant with a broad, if not total, generalization that renders the organism relatively flexible in its adaptabilities and capabilities, both mental and physical. Just as the human primate is the most generalized species in an order of primates that is the most generalized of mammals, so "our retardation relative to apes and other primates has been extensively documented . . . and the general retardation of development in primates versus other mammals—a phenomenon that is historically prior to the later differentiation within primates that establishes a tendency that is merely extended by the pronounced retardation of humans" (Gould 1977:366).

Generalization and retardation, manifest in neoteny, are the expression of progressive evolutionary strategies of mammals in general and primates in particular. This statement in broad terms may be contrasted to the strategies of other phyla and orders, which tend progressively towards specialization. The contrast is of course crude, but it is sufficient for our present purposes because the human species represents an extreme form of the strategy of generalization. The point of this contrast is this: that strategies of specialization fit more readily into the Darwinian theory of evolution by natural selection through adaptation than do strategies of generalization. Simply stated, when we talk of selection and adaptation, we can more easily suppose the selection of *something* differentiated and the adaptation of *something* to a definable condition—say, of a long neck to reach upper branches. But the selection of nothing in particular, the general, for adaptation to anything specific is, to say the least, rather difficult to comprehend. And though I confess that the puzzling nature of this formulation has more to do with semantic ambiguities than with the hard facts of the matter, the problem is best indicated in this way, as that which is at issue in the extreme case of the human species: why is human evolution marked by such an extreme degree of generalization, manifest in the extreme instance of neoteny and retardation?

The strategy of generalization enables organisms and their species to escape the constraints of specialization. The threat of extinction confronts the overspecialized, and adaptation by the generalized organism cannot become too specialized, for if it does, it contradicts the evolutionary strategy of the primate order. This, it seems to me, is the evolutionary condition *to* which human neoteny and retardation are adapted, and neoteny and retardation are selected for *through* this condition. In other words, any "instrumental" specialization of human morphology, mental or physical, would nullify the strategy of generalization and the advantages of freedom in time and space that it confers. Therefore selection must favor the regulation of instrumentalities necessary for the ongoing operation of the organism (brain, limbs, organs, digestion, and so forth) rather than improve the instruments to a specialized level. In this sense it is clear that the brain in particular has undergone a specialized development for reasons outlined earlier in this chapter, but not to the extent (yet) that it has overwhelmed our total somatic evolution as a generalized animal.

Thus there is a genetic outcome of the primate evolutionary strategy of generalization, as expressed in the human species, which favors the modification toward retardation of the genetic regulators of ontogeny. And the extrasomatic outcome of the evolutionary strategy is expressed in culture, the product of human thought (the brain) and human labor (the body).

The Logic of the Evidence and Its Conclusions

I have argued that paleontological evidence shows a primate and hominid tendency toward morphological generalization and enlargement of brain capacity. At the same time I maintain that the advent of human culture as the primary instrument of adaptive strategy cannot be accounted for by the evidence of, or inferences from, morphology. The generalization of the anatomy is associated with a development of the brain, but this development is unitary; the parts of the brain that develop are those concerned with the limbs and organs of the body, notably the motor and sensory capacities. The salient characteristic of human culture is that it is extrasomatic, and we therefore have to explain how it was possible for individuals, populations, and species to develop skills that enable them to control things outside their bodies, to use their bodies self-consciously as instruments to transform environments.

Since the evidence garnered from the material remains of hominid species does not seem sufficient to account for the development of the brain into mind, one obvious possibility is the proposition that the phenomenon resulted from evolutionary aspects of the "soft" parts of hominid existence, for which we can of course find no direct evidence. We can, however, examine indirect evidence, and providing that we establish the validity of certain premises, this indirect evidence may prove revealing.

If the morphological evidence from primates indicates an increasing tendency toward generalization and cerebral enlargement, then it is reasonable to assume that among the primates there is equally a "clinical" tendency toward generalization. For example—and I will discuss the matter in detail later—all primates (in fact, all species) provide distinctive evidence of patterned modes of organization relative to reproduction in particular. Orangutans are solitary animals; gibbons mate and grow up in a small "nuclear" family group; chimpanzees live in larger groups with a distinctive structure that is open enough to allow changes of membership. Taking the primate order as a whole, we find considerable variation, as the examples just cited indicate, and this variation testifies to the generalization characteristic of the order as a whole. If a mark of the hominids is extreme generalization, then we would expect this feature to apply in social matters. And if it does apply, then our attention would be drawn to the nature of the problems as well as to the advantages that such generalization occasions. And it does appear from the evidence that the human species has no fixed, biologically determined mode of social organization. Whereas ethologists can legitimately describe the chimpanzee as a group animal and mean this with reference to the entire species, there is no single apt description of the human mode of social organization. What, then, is the meaning and implication of this and other similar facts?

As will be seen from my example, the question addresses the primate order and humans as primates. The evidence that permits the question is taken from both contemporary observation of nonhuman primates and contemporary human social organization. In what sense, then, may this be related to prehistoric species of hominids, *H. habilis* and *H. erectus*, the ostensible subjects of the inquiry? Knowing that these species are ancestral to contemporary human beings, we can assert that all currently existing conditions of the human species have their origins in early hominids. The state of early hominids made possible the present, observable state of the human species. For example, we can say that if *H. erectus* is truly a member of the genus *Homo,* then just as the size of his brain is of human proportion, so too is the nature of his social organization. The creature's nature must be such that it makes possible social evolution, since we have direct evidence that human populations have devised social organizations that can be treated as objects analogous to tools: they are amenable to manipulation, elaboration, change, and transformation. We cannot claim that the social organization of *H. erectus* or early *H. sapiens* was one form rather than another—matriarchal or patriarchal, for example—but we can assert that *H. erectus* had a generalized potential for any of a number of possibilities. Which forms eventuated would depend on the circumstances and contexts of environmental problematics and on the relation between the different aspects of generalization of the species. At the same time, since there is no provable innate tendency to exist in one form of organization rather than another, our investigation must find or devise the actual instrumental means by which particular forms can be realized. We must imagine *H. erectus* or early *sapiens* not so much as a blank upon which any and all impressions can fall, but as a creature of immense but inchoate promise and potential, whose problem is to devise from his capacities the instruments and means to create specific modes of organization that will enable survival. This depends upon the discovery or invention of "artifacts," themselves of a sufficiently general nature that they can be adapted to the varying demands of specific environments. Just as Acheulean tools are specific artifacts of a general nature that enable their bearer to carry out similar functions in different environments, so I shall suggest that the invention of kinship is sufficiently general in nature that it can be shaped to produce different modes of organization that will fulfill necessary functions in different and varied circumstances. For reasons that I shall discuss in detail later, it seems to me that early hominids must have invented kinship, and that this response to hitherto undiscussed aspects of the environmental problematic is evidence of and contributory to the development of human culture based on rational thought directed extrasomatically.

To continue with the example of kinship and the nature of the evidence, symbolic kinship appears to be universal to the human species yet unknown to nonhuman primates (though many scholars have claimed to identify kinship among chimpanzees, for instance). If kinship is thus both species-specific and universal to the species, then we may consider it a diagnostic of that species on a par with any morphological features. If we consider kinship in an evolutionary light, the question is, What made this feature possible and what are its adaptive advantages? And further, What are its consequences? At a species level we may consider the presence of such universals as possible among prehistoric genera or members of a species, as well as among the historic members. If kinship was not a feature of *H. erectus,* then we have to say that at some early point in the career of the genus *Homo,* at least, kinship emerged to become a characteristic.

As this example suggests, the initial proposition is drawn from a comparison between human and nonhuman primates. The one has kinship, and the other does not. This does not mean, however, that I treat nonhuman primates as presenting a model of early hominid life. To the contrary, the whole point is that such a comparison reveals both common features and points of departure. The similarities I consider to be evidence of a heritage shared by all primates, which therefore cannot be considered instrumental in specifically human evolution. The differences lead our attention to the specifics of human evolution. This is where I differ from other theorists who speculate on human evolution. Whereas most other theorists search for similarities between ape and man and then adduce from those similarities a natural continuity, I maintain that wherever such continuities are in fact established, they cannot be considered *specific* traits of *human* evolution but are general features of primate evolution, which of course includes human evolution. If, for example, a theorist claims to discern "matrilineal kinship" in chimpanzees and takes this as evidence (1) for a natural origin of human kinship and (2) for a matriarchal original form, then I feel compelled to question this reasoning closely. In other words, if I find the matrilineal kinship of chimpanzees to be an established fact, then I shall have to admit that kinship has a direct natural basis and therefore cannot be used to account for the particular nature of human evolution. I maintain that the contributing factors to human evolution must be shown to have a foundation in the phylogeny of the species such that they do not appear miraculous in their advent, but at the same time it has to be demonstrated that in some way or other these factors are specific to the human species.

The major premise of my argument, then, is that as a primate species, the human is the most generalized not only in its morphology but also in its total inventory of dispositions and capacities. It is born to both uncertainty and promise. Whereas among other species we may find, for example, that the relation between the sexes for the purposes of reproduction is specified and particularly adaptive, in the case of humans we should find no determined and species-specific mode of relationship, but rather generalized features from which it is necessary to define specific modes. Therein lies not only man's adaptiveness but also his problem. By the same token, we may expect to find among the order Primates varying degrees and styles of diet and means of its procurement. Among human primates we would expect to find the largest variety of modes of extraction of food, as well as the broadest range of diet. In possibly the most crucial, yet at the same time indefinable, sense, the conditions of interaction between individuals of the human primate species will be the least well defined and biologically determined to occur among the order Primates. In terms of adaptation this means that it is possible for any individual to live with others, but most important, it means that an enormous problem has to be met by all individuals, namely, that little information about another individual can be known in advance and hence an individual has little advance information that will help him coexist with others on a predictable basis. Other primates, we may argue, can take for granted a far greater range of actions and reactions common to individuals than exists for the human primate. But if the human individual is to coexist with other such individuals, he must arrive at some ground for expectation and reciprocation. He must work out some common form of agreement about actions and reactions, one with some degree of reliability.

These and other necessities all derive from the phylogenetic tendency of the human primate toward a generalized nature that is the complement of the observed generalized morphology. But this generalized nature, the soft part of evolution, can only be rendered into actualities and specific forms by thought. No amount of brain activity directing the precision movements of the thumb and forefinger can produce an organization of relations between the sexes for the purposes of adaptive reproduction and nurture. The attention of the individual members of the species must thus be directed outside themselves and toward each other as complete, yet unknown, wholes.

To inquire into the development of this self-consciousness, subjective reasoning, and objective thought, we must search for the limits of primate specificity, for it is beyond those limits that the genus *Homo* has traveled.

2 THE PROMISING PRIMATE: Structure

Brainchild: The Meaning of a Complex Brain

We distinguish hominid fossils from other primate remains partly by the relative size of the braincase. As we move from *Australopithecus africanus* to *Homo habilis, Homo erectus,* and finally *Homo sapiens*, we have a creature whose probable brain size increases from 400 cubic centimeters to 1,500 cubic centimeters. That brain is housed in a cranium that becomes more and more vaulted, loses its ridges and crests, and shows more and more evidence of a forehead and a backhead.

The other primary criterion for distinguishing hominids from other primates is evidence of their upright posture and bipedal gait. The broadening and turning of the pelvis is particularly significant. With *Australopithecus* and *Homo sapiens,* the pelvis broadens and curves around to form a more boxlike structure that supports the trunk and from which suspend the locomotor muscles for the legs.

Together these two traits do more than anything else to define the anatomically human being. Together they are also contradictory developments. The modification of the pelvis for upright bipedalism results in a relative inflexibility to the bony structure surrounding the birth canal. The enlargement of the brain, and hence of its container, the cranium, means an increasingly large structure that must pass through the birth canal. As is well known, the difficulty is eliminated by bringing the human infant into the world before the brain and the skull have finished growing.

The result is the birth of an infant more helpless and dependent than is the offspring of any other primate and who remains so for a longer time. During this time of dependence, the brain completes its growth and under conditions that are outside the womb—in the environment of other people as well as in physical conditions. If we grant the generalized tendency of the human primate and the problems attendant on such generalization, it is not hard to perceive the adaptive advantages of brain growth *extra utero*. Still, a generalized creature must marshal its capacities so as to meet the specific and particular problems that confront him at birth. Extrauterine brain development facilitates adaptive learning about the ecological particulars that might almost appear instinctive in some aspects. This learning is, in a sense, a directing of the growing brain into certain adaptive patterns shaped according to the environment, and it constitutes a more or less natural marshaling of general capabilities. An obvious illustration is that all human beings develop the ability to speak language. The extrauterine growth of the brain means that individuals channel developing capacities into the effortless learning of a particular language. Presumably there is no universal natural language because the requisite areas of the brain do not develop to the full *in utero*.

The fact that the human infant is born prematurely, in a manner of speaking, has important correlates. The child is helpless at birth and is incapable of looking after itself for a number of years. It must therefore be watched by mature individuals until it can look after itself, if the reproductive process is to be successful. A. Jolly (1972:215) provides the following comparisons among primates for length of infancy:

Lemur	6 months
Macaque	18 months
Gibbon	2 years
Orangutan	3½ years
Chimpanzee	3 years
Humans	6 years.

Perhaps readers of Marx and Dickens and Kingsley would suggest that six-year-olds can do more for themselves than might here appear, but it is clear that the dependence of human infants on adults is greater than among other primates.

Since the helplessness of primate infants includes an inability to feed themselves as well as to digest externally provided food, the offspring is dependent on the adult female that can feed it, usually the mother. Nursing entails close bodily contact and is a two-way affair that offers the infant not only sustenance but also tactile satisfaction, warmth, and security. For the mother, nursing provides relief and also a sense of emotional satisfaction and closeness to the infant. In addition, since the human infant is immobile, it must be carried, and this again affords close contact between mother and child. The conditions of mammalian, primate, and human birth result in a close tie between adult female and infant. With the relatively greater dependence of the human infant, there is a relatively greater need for attachment, and we can assume from fossil evidence of early *Homo* that it was a feature of life.

At first the initiative for this attachment has to be taken by the adult, since the infant can only cry to attract attention. But by six weeks the human infant begins to smile and can begin to assume some of the initiative (see A. Jolly 1972:237). A smiling infant offers the mother rewards that tend toward a strengthening of the attachment. Indeed, it is possible to say that the mother and other adults seek the reward of the smile and, later, of other pleasing reactions from the infant. The relatively lengthy period of human infancy involves links with one or more adults. It can be said that the infant's behavior is regarded by those adults as being directed to them, for their sake, even though in an evolutionary sense the growing competence of the infant is primarily directed toward species survival. My point, however, is that the conditions of birth of the human baby and the protracted period of infancy make possible the development of an attachment between adult and child, a *primary bond*. This bond is reciprocal: not only is the infant dependent on the mother, but as this dependence progresses, the mother becomes dependent on the infant in several senses, notably for a degree of emotional satisfaction. Under certain circumstances, for example, when the adult becomes old and helpless, the positions between parent and offspring may be reversed, but the functions of the attachment may still be called forth (see P. J. Wilson 1978 for a discussion of further implications).

The primary bond of attachment in the human species is not something unique but represents a development of a common primate, indeed mammalian, trait. On its own it is not remarkable and may be considered natural rather than cultural, but as I shall show, when taken along with other traits, it can be seen to have important consequences. A bond between individuals is established over a lengthy period of time, to a degree of intensity, and covering a wide range of activity and emotion. This bond is firmly cemented long before the attainment of cerebral maturity by the infant, particularly the achievement of language, which permits the (continuing) attachment to assume a self-conscious status as something that can be talked about and represented.

It is notable that the primary bond pertains to adult females and infants. One cannot adduce biological evidence for such a tie between adult males and infants. In the primate order species vary widely with respect to the relation between adult male and infant. The marmoset male shows exaggerated paternal care, carrying around twin infants on his back until their combined weight nearly equals his own and handing them over to the mother only at feeding time (A. Jolly 1972:247–48). At the other extreme the hanuman langur almost completely ignores the infant. Between these extremes are many variations, from gibbon males, which remain with the mother and infants, to chimpanzee adult males, which cannot distinguish the infants that they sired. On the whole it can be said that most primate males are unable to identify their children (A. Jolly 1972:247). Therefore in ño sense can it be claimed that among most primate species, bonds or relations can be established between adult males and the offspring of females by reference to reproduction and/or nurturance. If connections are observed, then it may be assumed that they are determined according to other criteria, notably dominance/submission and "friendship" within a group based on residence. Particularly in the group-living species, it is true to say that such alliances are between males, and there is little bonding between males and females or between males and young. Even then, males in particular seem relatively free to wander between groups, giving further evidence of the temporary and loose nature of their relationships.

Between adult primates of opposite sex, the relationship is specific in its function but general in its structure; males act as protectors of the females and young. But there are variations. Among the hamadryas and gelada baboons, males live in a stable relationship with several females in what is called a harem group, but their protective stance toward the young seems to correspond to that of other primates (Kummer 1967:65).

Among nonhuman primates in general, there is a tendency toward specialization of the sexes. Females specialize in the rearing of the young and feed themselves in the process: males tend to specialize in the protection of females and young and in possible aggression against enemies and feed themselves in the process. Between females and young we can speak of relations determined by the biology of birth and nurturance. Between males there appear to be bonds based on dominance/submission, age, and proximity. Between adult males and females, there does not appear to be any form of particular relation between individuals; rather, all males live in a protective relation toward all females. This condition seems to differentiate the human from the nonhuman primate and is therefore worth examining further. I shall begin by surveying the range of nonhuman primate social organization.

Social Organization among the Hominoids

The white-handed gibbon is monogamous, and individuals that keep company with each other are an adult male, an adult female, and three or four of their offspring. Sometimes an aging male remains with this family. The male gibbon helps care for the young by grooming, playing with, and inspecting them. In some instances, male gibbons, in the absence of the mother, show considerable concern and solicitude for the young. Upon reaching adolescence, gibbons become aggressive and invariably leave their natal family to start families of their own. Gibbons live more or less confined to the forest canopy, though they may come to the ground occasionally (A. Jolly 1972; E. O. Wilson 1975).

The hamadryas baboon, which ranges across the acacia savannahs of Ethiopia, Somaliland, and Arabia, is terrestrial and lives in single-male harem groups, where one male possesses several females as mates. A number of these harem groups combine to form a band that provides protection during foraging. These bands also congregate on cliff ledges, where they sleep at night.

Young bachelor males sometimes form small groups of their own but gradually come to be accepted by dominant males as followers allowed to copulate with harem females until they can capture young females and eventually form groups of their own. A fairly close relationship is maintained between "master" and "apprentice" harem groups. Hamadryas females are relatively promiscuous in their matings (Kummer 1968, 1971).

The polygamous hamadryas baboon contrasts with other baboons, which live in large groups consisting of subgroupings of females and their offspring and centering around a hierarchy of males. These adult males are organized by mechanisms of dominance and submission, while the group as a whole orients its subsistence and protective activities by means of an attention structure in which subordinate males keep half an eye on the dominant males and take their cue from them. Mating is promiscuous, and the priority of access to females is determined by the status of males in their dominance hierarchy (Devore and Hall 1965, Chance and Jolly 1970).

The orangutan is arboreal and rarely comes to earth. It is also a rather solitary animal. A female and her young offspring remain together and may sometimes be accompanied by an adult male for a time. Lone males are common, and the group rarely exceeds more than four individuals, and then only for short periods. Male rivalry may sometimes occur, and an adult male may keep company with a receptive female for a while (MacKinnon 1974).

Gorillas live in isolated populations scattered across equatorial Africa and are the "one anthropoid ape species organized into age graded male troops" (E. O. Wilson 1975:535). The group is small, cohesive, calm, and low-keyed. A typical group might consist of one silver-backed or dominant male, a couple of black-backed juvenile males, a half-dozen females, and their young offspring. Lone males are common and may occasionally come together to form small groups. Gorilla groups occupy relatively stable ranges, but groups come into contact with each other without much drama. Gorilla life is said to be slow, and uneventful (Schaller 1965, E. O. Wilson 1975).

Chimpanzees are the one nonhuman primate living in an open society with groups averaging between thirty and eighty individuals. They occupy a fairly large but well-defined home range in the lower reaches of the forest. Unlike other primates, chimpanzee females as well as immature males may change their group affiliation. Mature males tend to remain in the one group.

The group includes a core group of females and their young, their grown daughters, and their offspring. Young adolescent males form a loose aggregate, or more accurately, a residual aggregate, but individual males do seem to retain a friendly, spasmodic contact with their mothers and sisters as well as brothers. The third segment of the chimpanzee group is the adult males, organized hierarchically through dominance and submission behaviors. Females mate promiscuously when in estrus, and to some extent the priority of access to receptive females is determined by the male dominance structure (Lawick-Goodall 1974, A. Jolly 1972, E. O. Wilson 1975).

Though by no means complete, this survey is enough to indicate the tremendous range of nonhuman primate organization, from the solitariness of the orangutan to the gregariousness of the chimpanzee; from the bourgeois monogamy of the gibbon through the polygamy of the baboon; from the closed nuclear family of the gibbon to the open associations of the baboon and chimpanzee; from male-female cooperation and equality among gibbons to separation and male dominance between the sexes among baboons, chimpanzees, and gorillas.

If we considered these examples as representative of the primate order, we would have to admit that social organization among primates is extremely generalized, for no one type is typical or can contribute toward the definition of a primate. Particular forms are typical of particular species, however, and all the examples cited are reminiscent of human social organizations—and that is precisely the point. One *species,* the human, lives in as many varieties of arrangement as an entire *order.* There is no single form that can be said to be typical among humans. So in the matter of social organization, considered in its most basic sense, the human species is to the primates as the primates are to the mammals: extremely generalized.

Although there is considerable variation among nonhuman primate species, each possesses a form of organization that is distinctive, permitting certain degrees of variation, true, but also limited. We are not likely to find monogamous gibbons under natural circumstances adopting baboon polygamy, or species that live in small groups, such as the gorilla, becoming gregarious like the chimpanzee, or the promiscuous chimpanzee assuming gibbonlike respectability. The variations that do occur are in group size, generation depth, and hierarchy. They are statistical in origin and nature, but they are not changes in type and form.

Since the human primate species cannot be defined by a distinctive social organization but manifestly appears capable of living not only in forms present among other primate species but also in some not found among other primates, it is extremely unlikely that this aspect of man's condition is in any way biologically determined. Human populations live in a wide variety of social organizations, suggesting that the human species might possess an innate propensity toward social rather than solitary living, and this propensity may be of such a generalized form that it can be manipulated to produce any of a number of possible variations. The attendant problem is to produce a specific *mode* of organization that is adaptive, and since there is no evidence for genetically determined general principles, the question is, How is it possible?

One answer has been suggested by Robin Fox who, on the basis of a more extensive comparison of primates than I have given, concludes: "There is little in our systems that is not found in 'nature' . . . , but nowhere is our system found in its entirety. The elements are common: the combination is unique" (1975:10–11). There is no question that the elements are found, but the model Fox presents is the traditional one. Somehow the human species put it all together, and naturally. Even if this is granted, the question still remains, How was it possible for the human species to combine the elements? or, Why could not some other species carry out the combinations, if these elements were there for some species to appropriate? But it is not a question of combination—where is there a human social organization that utilizes promiscuity as an element? And there are elements unique to humans—polyandry, for example. Furthermore, Fox's unnoticed shift from the plural "systems" to the singular "system" betrays the weakness of this argument—there is no human system, no means of organizing relationships common to the entire species. Characteristic of humans is that considered on a species-wide basis, their organization is generalized, that is, man lives according to a large number of specific types of social arrangement. Therefore it is highly unlikely that any specific form will be sufficiently generalized to allow the development of others. It is highly improbable, in other words, that the "original" hominid society was matriarchal, as many have claimed, or that it began in "group marriage," as others used to argue. Early hominids may have lived in a number of different possible forms of social organization because as hominids they were possessed of a sufficiently generalized principle and propensity to make any form possible, and further, their generalized condition made it mandatory that they shape from this potential something specific.

The problem, therefore, is not to guess the original form of social organization, but to work out the exact form of the generalized principle that would make possible the construction of the variety that we know human populations are capable of.

The clue that emerges from the comparison of nonhuman primate social organizations is two-pronged. We find certain species in which adult male and female, together with offspring, live jointly and relate more or less reciprocally. The gibbon is the primary example. But in such species there is no collectivity. On the other hand, among those species, such as the chimpanzee, where there is group life, there is no joint relation between male and female and no symmetrical relation of adult male/female to young. We can find the model of the human nuclear family among the gibbon, but no model of the group or of kinship. We can find a model of the group, and even of something resembling kinship, among chimpanzees, but nothing resembling the family. We can say that the human species is a "social" species, in that its individual members show definite preference, if not a need, for living in company. This tendency places man more with baboons and chimpanzees than with gibbons and orangutans. But this alignment does not provide a basis for concluding that human sociality represents an evolutionary progression from baboons and chimpanzees. We resemble gibbons in many other respects more closely than we resemble chimpanzees and baboons. We need to search for that which will account for the totality of similarities and differences with respect to social organization and will at the same time explain how the range and variation of man's social organization is possible.

The attempt must begin at the beginning.

Relativity: How Are Human Relations Possible?

The measure of successful evolution is successful reproduction. If we agree that the framework of human evolution is increasing generalization, the question now is, In what sense has human reproduction become generalized, relative to that of other primates?

The process begins with the copulation of two adults of opposite sex. In nonhuman primates, copulation is only engaged in when the female comes into estrus. At this time her body undergoes certain changes manifest in swellings, coloration, and odor. These serve as visual and olfactory signals to attract the male and wake him from his sexual dormancy. At the same time the mood of receptive females and attendant males changes. There are variations in estral patterning between species of primates, but of interest here are certain general features. The main point is that sexual receptivity, correlated with ovulation, is specialized among nonhuman primates in the sense that it is of limited duration and occurrence—most primate females are unlikely to be in estrus for as much as twenty weeks of a twenty-year life span (A. Jolly 1972:201). Similarly, the bodily changes are also indications of specialization, for once ovulation has ceased and receptivity diminished, the swellings, coloration, and smells disappear. When this happens, males and females do not copulate and show very little attraction to each other. Sexual activity among nonhuman primates is periodic and confined, and primate males cannot be said to have a sex drive in the sense that they maintain a constant interest or take the initiative in sexual activity. Their interest and arousal is conditioned by the physical changes undergone by the female. Although it is difficult for us to be certain, nonhuman primates do not appear to derive emotional satisfaction, pleasure, or disgust from intercourse. The act of copulation is completed in a few seconds and is not preceded by any foreplay, and although sexual activity is surrounded by male and female aggression as well as female moodiness, intercourse itself and the physical contact do not seem to be the source of sensual or emotional feeling.

A significant difference between the human and nonhuman primate is the loss of estrus among humans. Human females have a menstrual cycle that encompasses an ovulation period, at which time impregnation is most likely to lead to conception. But women are sexually receptive at all times (though less so some times) and can engage in intercourse at any time during the menstrual cycle and for most of the period of pregnancy. At the same time they show no accompanying specialized changes in body skin to signal the onset of the cycle. Human males, correspondingly, maintain a continuous interest in females and can engage with them in intercourse at any time during a span of sexual maturity lasting for fifty years or more. The human pattern of sexual receptivity is generalized, corresponding to the switch from specialized sexual changes, such as those found in nonhuman primates, to the permanent presence of elaborated secondary sexual characteristics, itself a generalized condition. People appear to gain pleasure and satisfaction from intercourse itself, at least far more so than do other primates, so that the act may be engaged in for its own sake. As yet no human society has been reported that does not make sexual relations a particular focus of interest. Sexual activity between the sexes among humans is subject to enormous variety. Unlike other primates, among which one single position predominates, humans engage in a large number of different positions and in a variety of tactile activities in addition to intercourse.

These phenomena are in keeping with the generalization of sexual behavior, which allows for the events of reproduction to be set in train at any time, without being subject to a determined specialization. The evolution of this generalization compares with the evolution of generalization in the human primate in other respects that we have noted and in still others that we shall discuss later on. As a result, the relationship between man and woman among humans is essentially defined by sexual activity. The receptivity of the female, the continued readiness of the male, the presence of permanent secondary sexual characteristics, and the complexity of activity all suggest that sexual behavior is a defining feature of the genus *Homo*. Ours is the only primate species that might be said to be "interested" in sex as well as in reproduction (the two being considered distinct).

The generalization of sex raises problems as do all forms of generalization. Since sexual relations are a permanent and generalized feature of species life, they must be defined specifically if sexual life is to be activated. If all females are receptive and all males ready, and if all males and females possess secondary characteristics that attract the opposite sex, then unless some means of control and selection is developed, there could be nothing but chaos. Chimpanzees, whose sex life is determinately periodic and controlled, can afford to be promiscuous, but permanent and continuous promiscuity under human conditions of sexuality is impossible. Sexual relations must be specifically determined from the generalized potentialities and capabilities of the species if successful reproduction is to be ensured. If all adults of both sexes living in proximity are capable of having sexual relations with each other at any time, then actual engagement by specific individuals at given times has to be determined from this general situation. It therefore seems likely that rules concerning sexual relationships were among the first to be devised by hominids as a necessary condition of their survival. We can assert without contradiction that all known human societies have such a body of rules and that such provisions have a fundamental importance.

Let us further consider the importance of the development of human sexual nature. Among nonhuman primates the relations between the sexes are not founded on sexual behavior. Among most primates, males and females live side by side but with a pronounced degree of disinterest or complementarity. They come together periodically, when females come into estrus, and in general males are related to females as protectors. Even in gibbon society, where a male and female adult live together, or among baboons, which live in harems, the "pair lives analytically, in opposition to others, as an independent unit, or within a community of similar units but without joint interaction with other pairs" (Reynolds 1968:219). Connections between individuals are based on primary bonding, on relations of dominance and subordination, on attention, and on proximity derived from territoriality.

But relations between human adult males and females are centered on sexual activity, by definition a joint activity. This focus is accompanied by a permanent and generalized basis for attraction and a necessity for specific definition. The actual form of activity and intercourse must be defined from the variety of possibilities attainable; the actual engagement between individuals must be specified from the variety of possibilities, and most important, the relation between sexual activity and access to reproduction has to be specified, for among human beings intercourse and reproduction become two separate facts. Sex, I have argued, is to some extent an end in itself, as well as being a means to reproduction. It has as its aim the definition of a *relation* between the sexes that is premised on innate differences. These differences are present in other primates but serve only to keep them apart, while among human beings the generalization of the differences brings them together, as individuals, into a relationship that can extend through time. Such extension need not, and may not, be permanent, that is, lasting the life span of individuals, but on the other hand, it is not the fleeting connection that we find among other primates or the connection established from other criteria, as among some primates.

In the human species, and probably in early hominids, there emerge two fundamental relationships founded on the generalized evolution of the species. These are the primary bond of attachment between an adult female and an infant and the bond of sexual attraction between adult male and female. The one is lengthy (relative to its counterpart among nonhuman primates), is intense, and becomes reciprocal; the other may or may not be constantly intense but has its moments and, more important, provides a basis for continuity that is absent among nonhuman primates.

I have tried to show that these bonds can be understood as natural evolutionary adaptations that are expressions (in the soft parts of human morphology) of generalization. Among nonhuman primates one can discern "foreshadowings" of the primary bond, in particular, such that its extension in humans and the implications of this extension make perfectly good sense from the point of view of adaptation. The sexual bond of attraction, though not as explicitly foreshadowed among nonhuman primates, is a necessary consequence and part of the generalization of the sexual organization of the species for reproduction. Both bonds belong to, or are different aspects of, the same process, namely the successful reproduction of the species.

Among nonhuman primates the various stages of the reproductive and nurturing processes are independent in the sense that copulation and conception are specialized, being confined to estrus, while birth and nurturance are specialized, being confined to females. In some primate species the male plays a direct role in the nurturance of the young, but this role is specialized, determined. But with the development of the primary and pair bonds in the human species, that which remained separate and specialized among nonhuman primates becomes related and generalized. And whereas among nonhuman primates the adult male and female remain separate, in the human primate it becomes possible for a bond to form between them.

The human extension of the primary bond, together with the continuation and interdependence of that connecting the pair, creates a general situation whereby it now becomes possible for the two liaisons to overlap, to interact, and thereby to produce the possibility of a third relationship—that between adult male and infant.

This third relationship differs from the other two in that it is mediate, the product of the conjunction of the other two, which are immediately generated by the biological conditions of reproduction, nurture, and attraction. The relationship between an adult male and infant is possible only through their common connection with the adult female. The relation is premised not on the biological role of the male in the conception of the infant but only on the continuity of his relation to the female, which must be sufficient to overlap with her involvement in the primary bond. It is certainly more than likely that in the majority of instances the pair-bonded male is the genitor of the infant, but it is not necessary that he be so, and given that cultures exist even today that either do not recognize physiological paternity or give it no place in their ideas of reproduction, it is important that we recognize the real basis for the relationship. There have been long debates over the determination of kinship by biology.

Once the possibility of this new form of relationship is actualized, or if the connection between the two major bonds is identified, then the adult male, female, and young are related together according to principles independent of sex linkage, sex difference, and generation difference. Among other primates, principles of organization through relationships are founded on the separation of the sexes (as among chimpanzees) or are defined in such a way as to constitute a specialized form of organization (as among the gibbons). Among chimpanzees, adult males are organized separately from females and from young males: among the gibbons, male and female adults are pair bonded, but inflexibly so, and no means of transferring relationships exists. Yet the human situation makes possible changing relationships between individuals according to the general principle implicit in and initiated by the relation between primary and pair bonds, or what is the same thing, the emergence of the father relationship.

If it is possible to create a relationship by combining two others, then the principle of transitivity emerges, by which a link between two individuals may be transferred to a third. If relations between individuals do not have to depend on biological determinism, they can become generalized. Specifically, if it is possible to create the relationship of father by conjoining the pair and primary bonds, then since sex and generation boundaries have been crossed, there are no barriers left. The relations between any two individuals can be continued indefinitely by linking through the common member(s). The extension of the primary bond and the lengthening of the pair bond lead to a junction that is the elementary basis for a generalized social organization typical of the human species. For it is only then that kinship is possible, and with kinship we have the most flexible, generalized, and adaptable principle of group organization in the primate order. This principle provides the species with the means to organize groups of any (or almost any) size, from a minimal nuclear family to a maximal tribe, and according to any one of the number of possible specific forms. From the psychological point of view, the genesis of kinship engages the male in the entire process of reproduction and nurturance *directly,* not just indirectly, as one among many protectors of the females and infants as a group. The adult male is retained in the reproductive/nurturance cycle and the specialized primate adaptations, which depend on the division and specialization of the sexes and lead to the organization of groups according to separate principles, are overcome. This elementary kinship thus permits humans to remedy the failure of the gibbons' type of organization to allow group formation and the failure of the chimpanzees' type of organization to integrate male, female, and young.

At this point in the argument, having introduced the term "kinship," we must examine the argument relative to other hypotheses before developing further the implications of the present theory.

The outstanding feature of human kinship is that it unites both sexes and all generations by focusing them on reproduction and in this way creates both individualized relationships and a generalized possibility for the formation of groups. In this sense the emergence of kinship as the means of organizing people socially can be understood as in keeping with the evolutionary tendency of the hominid species toward generalization. Social organization is certainly a major adaptive dimension, and the supposition that a particular form originally existed would simply be out of keeping with the pattern of hominid evolution. A social organization founded on kinship is generalized and so of necessity must be defined specifically in any one of a number of possible forms. Which forms in which case depends on the perceived, or worked out, adaptability of possible forms. Early hominids, then, knew or invented kinship but must have lived in social organizations that varied according to environmental circumstances— some may have lived in "bilateral" groups, others in "unilateral" groups, some may have recognized linearity, others may not. The most we can say is that they acknowledged kinship.

And this is precisely what cannot be said of nonhuman primates. My argument here is in direct conflict with those of nineteenth-century thinkers, who claimed that matriarchy was the original form of human social organization. Such arguments still hold sway and have recently been revived in a modern guise by no less a scholar than Robin Fox, whose reasoning I will dispute on its own terms.

Fox suggests that nonhuman primate groups are built of elementary "blocks" composed of adult females and their young, adult males, and juvenile males. These groups are said to form two basic systems: the single male and the multimale. Such typing according to male presence is misleading, for it suggests that the group hinges on the male. This may be true in some cases, but not in all and not among chimpanzees, which Fox takes as his archetypal example. For among chimpanzees, the core of the *group* around which others revolve is the line of females and their offspring, and the structural foundation of chimpanzee groups may therefore be said to be female.

These primary blocks of females "are basically units of uterine kin—or not to put too fine a point on it, matri-lineages" (Fox 1975:15). Apart from the obvious criticism, that the terms "uterine kin" and "matrilineages" refer to human organization and hence are at best analogous, having by no means been proven homologous, the question is how far chimpanzee matrilineages in fact resemble those of humans. For the chimpanzee matrilineages consist of females, only temporarily including immature males. No human matrilineage is so constituted, for all include males—

young, adolescent, and mature. All human kinship systems entail males. Even if we did grant that chimpanzee females form "matrilineages" and hence provide us with an original form of human organization, we would have to explain how it became possible for the original practitioners of this system to incorporate adult males—and to do so by transforming the identity of adult male to *father.* Since chimpanzee mating is promiscuous, there is little or no possibility that a chimpanzee male will lay claim to the offspring of a female for reasons having to do with reproduction. Since sexual attraction is temporary, periodic, and obviously determinate, there is no possibility that a chimpanzee adult male and a female will establish a pair bond sufficiently coherent for transferral to the primary bond. Neither Fox nor anyone else has even recognized this problem, let alone attempted to answer it. But since chimpanzee matrilines do resemble human organization to some extent, and since we have dismissed this resemblance as based on incipient kinship, how else can it be explained?

The chimpanzee primary bond is, as we have already noted, almost as lengthy and intense as that among humans. There is clear evidence that mother and offspring enter into a close emotional as well as physical relationship. It is equally plain that this liaison is intense enough and long enough, and that chimpanzees are aware enough, for it to be an individualized bond. The tie between chimpanzee mother and daughter is sufficiently well established emotionally to outlast the need for physical, nurturing dependency and to overlap temporally with the establishment of a new primary bond between the daughter and her offspring, as well as between the several offspring of a single mother (hence defining even a tenuous, but nevertheless real, link between aunt and niece).

Chimpanzee matrilines then, are the product of extensive, overlapping primary bonds reinforced by the continuing proximity of component females in the same territory and group. The line forms the stable core of a group that is otherwise open, permitting membership to change. Such a society no doubt allows for relatively flexible adaptation, but the female line at best only simulates human matrilineages. The problem, unrecognized and unaccounted for, is "chercher l'homme."

The chimpanzee male remains outside the pale of relations founded on reproduction, and the chimpanzee "community" is organized, as it must be, on principles specialized according to age and sex. But by transforming the adult male, who merely impregnates the female, into an identity we can term "father," the human species abolishes determined social specialization and creates a generalized situation in which male and female and all generations can be interrelated, through kinship, in one of a number of possible ways.

Fatherhood and Kinship

I argue that human kinship is made possible with the transformation of the adult male into the father. This assigns to the male a position complementary to that of the female in the reproductive process and unites both, according to a principle in which they can share but compounded of qualities that permit extensive variation. The most important of these features is transitivity, the transfer of relationships.

The initial link between father and child is created by passing to the child the male's relationship with the female, thus establishing the general principal that relationships may be transmitted. Although there is evidence of such transmission between chimpanzee females, we have seen that it cannot apply to males. Although gibbons form permanent relationships, there is no evidence of their transitivity. A nonhuman primate organization of males, based on dominance and submission, has to be renewed with each generation. It has been suggested that chimpanzees (sometimes) and Japanese macaques (more often) are able to transmit status through females to males. But since juvenile male chimpanzees leave the troop when dominant males are at their height, the transfer cannot apply to them; transmission of status through females could only occur when dominant males are declining and juvenile males left in the troop have a chance of gaining the ascendancy. There seems to be no evidence for the self-conscious application of transferral of relation as a principle of status.

In want to emphasize—and will therefore repeat—that the "invention" of the father is of necessity founded not on the biological facts of paternity but on the relation of a male to a female and on her relation to her offspring. The term "father," as a kinship term, is best understood as meaning "mother's mate/husband," a definition found in many kinship terminologies. It is a cultural relationship, a derived relationship. It can be formed only if (1) a male exists in a pair relationship with a female and (2) that female has offspring.

The identity and relation of father is thus the primitive or elementary social relationship, constructed by transforming natural circumstances and "materials." This view suggests an explanation for the "male domination" of society. Human culture, particularly as social organization in general and kinship in particular, only comes into being when the adult male is transformed into a social father. In culture it may be expected that "male" principles predominate. This is not to say that men are or must be dominant over women, but rather that human culture has a bias toward the male idiom because of the terms of its development.

Still, this evolution, it will be appreciated, depends fundamentally on the female, who stands at the intersection of the primary and pair bonds that make kinship possible. The male's identity as father relies on the female not only logically and structurally but also emotionally. For it is the female who "admits" the male to the pair bond and thence to the primary bond and who, in many cultures, is literally free to designate the "father" of her children, thereby determining the identity of the male. Similarly, in many cultures it is notable that unless a man is so transformed, he dismisses his wife (and in some societies he may reclaim payments made before marriage).

These examples, of which the prime instances are to be found in the African institution of bridewealth, all testify to the frailty of fatherhood, to the fact that failure to achieve it may cause the kinship system, hence the social foundation of the community, to fail. If "father" were a biologically determined identity, then it would not need constant reinforcement and enlargement. But because fatherhood and kinship are the first cultural facts and are entirely dependent on females, the male position is precarious and the structure of society fragile. Male cultural dominance is a function of male vulnerability, which must be protected. Thus although females give birth to children, in the majority of human societies it is males who "claim" them and control them, not so much as individuals but as productive, reproductive and social resources.

Such dominance is cultural, social, idiomatic, and theoretical and should not be thought of as physical. Males do not dominate because they are physically stronger, though they may do so in individual instances. But such domination cannot occur as a general principle of culture, since the human species would end in Hobbes's war of all against all—and for Hobbes, that is where we supposedly began. Nor can male dominance derive from a sexual division of labor in which man the hunter prevails over woman the gatherer. While human hunters and gatherers may prefer to eat meat, there is no evidence, except in extreme environmental situations such as those of the Arctic Eskimo, that human survival depends on hunting and hence on male activity. Rather, the reverse is true: gathering produces by far the greater proportion of the subsistence of hunters and gatherers. Godelier (n. p.) has suggested that with hunting, males become oriented toward death and females toward birth, that hunting males become accustomed to the use of physical force, and hence that there is a material basis for male cultural dominance. But this argument again leads to the Hobbesian state and cannot be valid. In any event, if we want to take the case literally, a woman is also capable of handling a tool as a weapon, and since the human species is a thinking species (and thinking involves no sexual discrimination), women can compensate for their physical weakness by cunning. The argument for male domination through the greater utility of male economic activities is weak not only for hunting and gathering societies but for pastoralists, agriculturalists, and horticulturalists, and may well become so for industrialists as well.

The crux of the explanation of male cultural dominance is (1) the male dependence on the female for the genesis and continuation of culture, particularly social organization founded on kinship, and (2) the fact that the center of the transformation from "nature" to "culture" is in the transformation of the adult male (nature) into father (culture). This fragile weakness must be strengthened not by any natural forces (such as physical strength) but by cultural factors in which male fathers continuously impose and reinforce their position and hence the culture and social structure of the given population. I shall suggest later that the primary ways in which this is done may be subsumed under the heading of ritual, where male cultural dominance is created. Females are not necessarily excluded (though they may be, from male rituals), but in the domain of ritual activity, male dominance is the focus.

Before discussing this further, I wish to specify the conditions of kinship. For while it is one matter to outline the prerequisites, it is another to specify the instruments that translate kinship from an inert possibility into reality. If we are to understand the evolution of the human species through culture, we must attempt to explore this question.

I suggested above that for the human species, sexual intercourse is the behavioral act that defines the relation between the sexes. To put it bluntly, male and female, as species beings, are characterized by their intense, continuing, mutual sexual interest in each other. In this respect they may be contrasted with other primates, for which sex is a periodic activity of only occasional concern and the mutual relationship between the sexes tends to revolve around protection. Among primates, the human species alone is capable of continuous sexual activity, hence of a continuous concern with sex. Whereas sexual activity in other primates is specialized, hence confined and determined, among human beings it is generalized and undetermined. If sexual activity is to occur in the human species at all, then the time, place, and persons have to be specified, defined from the general circumstances. It might also be added that since human sex is physiologically generalized, there is no specific way that an individual instinctively knows how to have intercourse, and the procedure has to be learned.

It is the very fact that sex is the defining feature of the relation between the sexes that makes *control* and *definition* of sex necessary if intercourse is to take place and lead particular individuals into pair bonds. How is this possible? How can the generalized nature and potential of the species ever allow individuals to be sexually attracted?

The major problem for the evolving hominids is to figure out how to activate their bodies in particular ways so as best to exploit the resources of their environments. By the same token the social aspect of generalization poses the problem of how to define specific modes of organization for particular survival activities, including reproduction—which in turn involves sexual interaction. Thus the evolving hominid has increasingly to think about himself and, in the matter of sex, to think how to select and define specific relationships. Precisely because this is man's situation, a cardinal human trait is an enlarged and complex brain; as the range and intensity of the human problem evolves, as human beings can take for granted less and less about each other, the powers of the brain for thought and communication are challenged and developed.

The possibilities of pair and primary bonding are converted to actualities by the imposition of rules that define and specify relationships in particular, and these rules apply most immediately to sexual intercourse. If sex is the foundation of relations between males and females in general, then any actual relationship cannot be founded on uncontrolled sexual activity. Real relations between real individuals have to be determined by regulation or definition, which in turn makes possible the formation of specific social organizations. Since there can be no built-in, instinctive mechanism for the purpose (if there were, then sex would be specialized), regulation has to be contrived and implemented by individuals among themselves. The human populations of the species must make themselves the object of their own thoughts and subject to their own, humanly devised procedures. While theoretically or in the natural state, all members of the opposite sex are attracted to each other and are capable of engaging in sexual relationships, in practice only some members of the opposite sex are permitted to do so. If we recognize that this is so, then the possibility of identifying the adult male as father and of making relationships transferrable can be realized. Then kinship becomes possible, and there exists a foundation for the creation of varied but specific forms of group social organization that can form the basis for the structuring of cooperative and competitive survival activities. Since it is unlikely that the early hominids (unlike the nonhuman primates) instinctively possessed any determined organization making cooperation possible, the need for one would be a matter of utmost urgency, certainly a matter for immediate experiment.

The Incest Taboo

We should note at this juncture two related implications of the generalized adaptive pattern of the human species. (1) Sexual adaptation is extremely general, therefore individuals and populations are free to adapt to a wide range of environments, but in order to do so they must form an adaptive pattern to suit their environmental situation. (2) Though the human species is innately a species of unsocial sociability, in Kant's phrase, individuals and populations have to work out specific forms of social life for each and every situation in which they find themselves.

In both instances any particular pattern adopted may be built on natural conditions but cannot itself be one. Second, since some sort of transformation of natural dispositions is necessary, some one or more principles are necessary to make this transformation possible. Adaptive efficiency dictates that they must themselves be sufficiently generalized to permit the formation of specific adaptive patterns.

To illustrate, according to the argument thus far, pair and primary bonds are biologically based. Their natural co-occurrence provides the possibility for creating a new relation, the father, which itself contains the possibility for kinship. The actualization of these possibilities depends on their development by the species, and we assume that the early hominid brain was equal to this. Once kinship and fatherhood have been established as facts, their utility can be exploited. Kinship's adaptive advantage is that it provides a generalized basis for specifying sexual relations, individual identification, and group organizations. But such specification remains only a possibility unless and until further principles are formulated that establish kinship as a reality and as a universal. This is particularly true in the sense that once the possibilities of kinship have been realized, then their achievement must be preserved by institutionalizing kinship, on the one hand, and ensuring its continued reproduction, on the other.

Suppose we take a simple hypothetical model in which time stops. Our primordial individuals discover that a bond can exist between the adult male and the offspring of the female and that the link provides a principle, enabling them to relate individuals together, irrespective of sex and generation. This grouping confers advantages of assured sexual access and effective nurturing. Such a group may also prove economically efficient as well. But if this arrangement is not to prove a "once only" matter, then both the principle of kinship and the place of individuals vis-à-vis each other must be established so that it may be perpetuated. If the crux of the matter is to ensure that every male on reaching adulthood is transformed into a father, that is, into a male whose position with respect to his offspring is central rather than peripheral, then operative principles are necessary to avoid the need for rediscovery of kinship by every new generation of *Homo sapiens*. Although principles are in a certain sense independent of kinship itself, they are necessary for the establishment of kinship in the life of the species.

It is important for us to recall that kinship derives from the role of the father and that being an original cultural factor, it has no guarantee other than that which can be imposed by the will of populations. Simply stated, an adult female will be naturally transformed into a social mother when she bears a child, but there is no corresponding natural transformation for a male. If he becomes a social father he does so only culturally or symbolically and, as it were, by permission of the female. Yet for the continued existence of kinship and the structures founded on kinship, it is essential that the concept of the father be perpetuated. Only if we bear in mind this meaning of "father" and the vulnerability of the father to nature and the female can we explain the dominance of the male/father in all human cultures, which can first be seen in the asymmetry between the relations of mother/offspring and father/offspring.

Many human societies, including our own, have the concept of "bastard"—a child without a father, an "illegitimate" child. No child can be illegitimate through the mother, but a woman who bears a child and fails to admit an adult male to fatherhood is blamed. It is invariably the woman who is stigmatized or punished for bearing a child "out of wedlock," because by doing so she undermines the institution of kinship itself, though in individual cases this consideration may not be uppermost. In any event, when this occurs, the condemnation is public and often is made by members of the community who are not affected personally. They condemn "on principle," and the individual is shamed. The asymmetry between male and female relations is similarly apparent in adultery. Even in the Trobriand Islands, where there is a matrilineal emphasis and a ritual diminution of physiological paternity, a woman who commits adultery is punished more severely than a man—if not physically so, then by shaming. Such asymmetry and "unfairness" are simply the protection of the male's vulnerable position toward the female and in the long run defend the precarious cultural principle of kinship itself.

If more evidence is needed to support this general contention, one need only consider the varieties of cultural practices focused on establishing parentage, particularly fatherhood. In Caribbean societies men and women are free to engage in sexual relations without at any time being compelled to enter permanent, monogamous union, though many individuals do. Many males boast of their sexual prowess, but their very standing as males is confirmed not by such boasts but by proof of their deeds—the children they have sired. Women are as free as men to engage in intercourse and as independent, and it is up to them to designate the father of their child, which they do with quite deliberate cognizance of the power they wield in doing so (P. J. Wilson 1973). At the other extreme, the example of African bridewealth confirms the asymmetry and the vulnerability of the male. The male at marriage hands over cattle and goods contributed by members of his kinship group. If no children are born, a divorce may follow, in which case the bridewealth is handed back. If some children are born and a divorce follows, some cattle are handed back. Among the Tsimihety of Madagascar an increasingly popular practice is the trial marriage in which only a token bridewealth is handed over at marriage, and if a child is born in a year, further payments are then made (P. J. Wilson 1967).

Yet another widespread cultural factor reflects the general principle of male/female asymmetry. In many cultures a male fails to become a full person until he has children. In some societies, such as the Nuer of the Sudan, if a man dies without issue, a woman may be married to his name and her children may be reckoned to be his children, for only then can his life be considered fulfilled. Only if all males can be transformed into fathers can the kinship system, and the society built upon it, sustain and reproduce themselves.

Bearing this in mind, let us now turn to the incest taboo. This prohibition forbids sexual relations between kin. People must be identified as kinsmen *before* sexual relations between them can be prohibited, hence kinship is logically independent of the incest taboo, but the taboo depends on kinship. The prohibition emerges as a positive, existential fact only following indulgence in the forbidden sexual relations and only when a third party finds out (or incest is publicly admitted). Under these conditions alone will the sanctions that define the taboo become a fact of life. True, one who commits incest may feel guilty, even though the secret is never divulged—but guilt is a social phenomenon engendered by having to bear the burden of keeping secret what one knows is condemned. Being found guilty is, more accurately, being found out guilty. It is interesting to reflect on the characterization of incest and other forms of sexual activity (such as bestiality) as being "unnatural," when in fact such acts, and others, are natural but "uncultural." A nature red in tooth and claw "commits" murder, robbery, and incest and does not, cannot, condemn them. It is precisely because such acts are natural that human society denounces them or taboos them. And this is because their commission threatens the culture of a community at its very roots, where it defines its humanity.

So whether the incest taboo, or more vaguely, incest avoidance, is natural to human beings is really beside the point. If it can be shown to be intrinsic to human biology, we must nevertheless still ask why and how a natural avoidance assumes specific, variable cultural form and relevance. Why and how has it been transformed from a neutral natural fact to a stringent taboo imposed by others?

Suppose that we consider the evidence from the kibbutz, where unrelated young children of both sexes grow up together and avoid sexual relations with each other, or at least marriage. This situation shows that human beings who grow up together avoid sexual mating. Let us suppose that the kibbutz offers a paradigm for the original conditions of human existence. Now, the human incest taboo specifies that persons who are brother and sister (kin) must not have sexual relations. Can we explain how the paradigm—the rule against relations between persons growing up in close proximity—came to be transformed into a prohibition against sex between brother and sister (or mother and son or father and daughter)? Or, why are a brother and sister who grow up in *separate* kibbutzim not *allowed* to have sex or marry (though they might wish to)?

What individuals think and do is not the issue, at least not directly. Nor is it a case where genes seek their own best interests by directing individuals. Rather, principles of culture preserve and reproduce themselves through human beings. Of course, some being called "culture" does not literally direct matters. It is simply true that since kinship exists as the most flexible and adaptive principle of organization, human beings must adapt their activities and behavior to its conditions.

The incest taboo is that general principle by which kinship perpetuates and reproduces itself. It cannot be responsible for the origin of kinship as prevailing theory contends. Specifically, this prohibition makes possible the continued reproduction of the relation "father" by ensuring that the natural preconditions of kinship, a pair and a primary bond, remain separate and integral. The rule against incest institutionalizes or culturalizes these bonds, defining and guarding them as preconditions of kinship, protecting them as vital natural resources. Without the incest taboo, the bonds and their conjunction would remain natural facts whose possibilities were unrealized and unexploited.

The prohibition as a cultural rule is not concerned with individuals, their psychology, or their biology. It may be that individuals are tempted toward incest, or it may be that they have a natural aversion to it. Such inclinations make the rule either harder to observe or easier. If some or all people show the former inclination, then the imposition of the prohibition is more than likely to have psychological consequences, and it was upon this possibility that Freud based his theory of the incest taboo. If the aversion is more natural, then the taboo simply reinforces it and is somewhat redundant. In this case we should adduce some reason for the species' continued imposition of the incest taboo, since its own nature automatically avoids incest. So whether the incest taboo is proved innate or not, a question remains and an answer must be looked for.

The most severely condemned form of incest is that between mother and son. Many societies consider this form unthinkable, and statistically it is the most infrequent. But there is no doubt that it is possible. As the extreme infraction of the taboo, it may constitute the most extreme negation of the conditions of kinship. In fact, a sexual relation within the primary bond makes a separate pair bond impossible and fails to transform the male into father, since he has already been defined as the dependent party of the primary bond. Mother/son incest, if sanctioned in principle, simply reinforces the natural dominance of the female by retaining the primary bond as the only form of relation. The male cannot enter into an autonomous relation of kinship with the offspring of the female, hence cannot become father (though he may have sired offspring). Mother/son incest would not make physical reproduction impossible, though if it continued, sibling incest would also have to be permitted.

The brother/sister incest taboo is regarded as the most significant of incest taboos in Lévi-Strauss's (1967) theory of kinship. He argues that because it prohibits sexual access between a male and female most accessible to each other, it compels them to find partners elsewhere. Since a man cannot be expected to give up rights to his sister without expectation of rights to another female, Lévi-Strauss concludes that kinship and marriage begin in the exchange of sisters by brothers. One reason why I cannot accept this is that the incest taboo can only be defined by kinship, incest meaning sexual relations between kin; therefore the incest taboo presupposes kinship, yet Lévi-Strauss is claiming that the incest taboo gives rise to kinship. Incest taboos can be imposed only when kinship has been created and the conditions for this are the delineation and perpetuation of the primary and pair bonds. Brother/sister incest would not allow this.

Brother/sister incest allows the creation of a new primary bond, but precludes the formation of a new pair bond, since the male and female mate *because* they share child status in a primary bond. Even if they were attracted to one another, that attraction is preempted by their preexisting primary bond relationship. Their relationship may gain an added dimension, but it will not be a new relationship. The man remains brother and at best comes to share his primary bond relation to his sister with her child. They become a single, almost indivisible unit, and presumably this exclusiveness was the point of brother/sister incest in such royal dynasties as those of ancient Egypt and Hawaii.

Under brother/sister incest a man cannot achieve a standing of his own, that of father, since his connection to a child is subordinate to and enclosed with his already existing primary bond relation to his sister. Brother/sister incest does not allow a man to become father hence would not permit the establishment of kinship. The taboo on sexual relations between members of the same primary bond contributes to the maintenance of conditions that make kinship possible.

Whereas mother/son incest is frequently unthinkable, sibling incest is generally thinkable but severely punished, particularly in matrilineal societies, where the natural, pivotal position of the female is recognized explicitly. Women collectively are the sole agents for the reproduction of individuals, and so the continuity of life itself is assured. Furthermore, following the birth of children, women automatically mature in status. The male in such a system remains entirely dependent on the female; only if the male in matrilineal societies is *freed* of his shared primary bond to his sister can he have the chance to achieve an autonomous position in relation to succeeding generations, a position in principle equal to that of the female. Thus in matrilineal societies sexual bonds must be kept separate from primary bonds to allow the male to become a social father and the society to construct a kinship system.

Father/daughter relations are the form of incest least condemned by most human societies and statistically the most frequent. Father and daughter do not exist in a primary bond but are related solely as kin. Their sexual liaison therefore will not conflate primary and pair bonds, as do mother/son and sibling incest, and does not constitute a threat in principle to the possibilities of kinship. Through father/daughter incest a new primary bond is created, but a kinship relation is conflated with a pair relation, resulting in what we might call a sterilization of kinship. The male is already transformed and is father, so that by establishing a pair bond with his child, he simply remains father. Similarly, although the daughter becomes mother, she also becomes sister to her own child, and in this situation a shared primary bond of sibling is conflated with a primary bond of mother. The generative possibilities of kinship are closed off with father/daughter incest, because the autonomy of the pair bond is eradicated, thus removing one of the two preconditions for the existence and reproduction of kinship. This idea permits us to reinterpret Freud's theory of the incest taboo as it pertains to the Oedipus complex, where the monopoly of daughters by the father excludes the sons from sexual partners, thereby rendering them sterile and undeveloped males who cannot become fathers.

Any discussion of the incest taboo can only be theoretical, by which I mean it can only treat principles. Given the generalized sexual nature of the human species, we can hardly assume a natural disposition to mate *only* with proximal members of the opposite sex. The prohibition, however, is aimed at the possible existence of such a state of affairs. Its universality and stringency can only be explained if it is shown to be concerned with principles and not with individual cases. That is, the incest taboo condemns individual cases only because if incest became universal or generalized, then kinship and the control of sexual relations would not be possible. Individual cases of incest do take place and do not of themselves jeopardize a kinship system, nor even physical reproduction. Actual occurrence testifies to the lack of a universal abhorrence of incest, in the sense that at least some individuals are not inhibited. But individual cases are condemned by others, if they come to light, because their occurrence contravenes a social or cultural principle, the logical conclusion of which is: if everyone were allowed to commit incest, then kinship as we know it could not continue to exist.

The present argument is in some ways analogous to that of Freud, except that my contention is that the incest taboo has consequences that are destructive socially rather than, or as well as, psychologically. For Freud, permanently incestuous relationships would mean intense rivalry between unequal competitors for sexual access to the members of the opposite sex: father and son would compete for mother/daughter; mother and daughter would compete for father/son. As it is, the suppression of such desires by the taboo is the cause of forms of neurosis. I believe that this may be so, but I am most concerned with demonstrating that if incest were generally permitted, then the principle of kinship as we know it would not be possible and could not be taken up as *the* adaptive principle for the construction of social organizations. I do not mean that some other general mode of social organization might not be forthcoming, but if it had been a historical possibility, we would surely have the evidence for it. The facts are that we know kinship to be universal in the human species, yet absent from nonhuman primates, and that we know the incest taboo to be universal and to be couched in terms of kinship. Kinship, however, exists only in specific variants, as systems, so the single universal form of it can be said to be meaningful only in principle. The statement of the incest taboo in terms of kinship in general is likewise a matter of principle that will have to be specifically stated for each system. Any general theory of the incest taboo must refer to principles of kinship and cannot be founded on instincts or on any universal physical fact, for this prohibition is a logical principle, not a biological one.

The incest taboo focuses on the primary bond and is the means whereby the primary bond is assured its integrity. But the incest taboo treats the pair bond residually. While the prohibition extends to certain possible forms of mating, it does nothing to prescribe, or even encourage, others, nor does it contribute to any assurance of the pair bond. Yet if kinship is to be possible, it is as necessary that the pair bond retain a degree of integrity as that the primary bond do so.

Marriage

The principle of marriage is as universal to the human species as kinship and the incest taboo, and this in itself may be taken as indicative of the fact that all three institutions are parts of one whole. By marriage I refer to some mode whereby a pair bond between adult male and female is transformed from a natural attraction to a cultural fact. The crucial fact is that this relation is identified or objectified by deliberate, creative human action, notably ritual. This need not involve elaborate ceremonial, but it does entail some sort of public marking, generally an exchange accompanied by public recognition that the children of a particular woman will recognize one man rather than another as "father." As with the incest taboo, marriage emerges as a social fact in the sense that other people are enjoined to uphold the relationship and to assume responsibility for exercising sanctions or for supporting their exercise. The institutionalization of the pair bond through marriage imposes restrictions on the sexual relationships of the individuals concerned. Usually these are more strict for the female than for the male, since it is the male's title that must be protected more than the female's.

In those societies where marriage is formally ritualized, the pair bond is instituted before the birth of children, which may be said to confirm the situation of the male that was anticipated by the marriage. But there are many societies in which the preliminary establishment of a pair bond is absent or minimal. In Caribbean societies, for example, people frequently are not ritually married until they are past childbearing age. The most common form of cohabitation is common-law union, in which an adult male and female simply live together and mutually trust each other for common provision. Although a certain demand is made by the male for exclusive rights to the sexuality of the female, it cannot be enforced in such circumstances, but so long as the male lives in a de facto "husband" relationship, he may claim the right of fatherhood. Where no stable relations exist, the woman has the right to designate the father of her child, whether or not the man named sired the offspring. A man so designated in most instances is only too pleased to accept his title and will from time to time confirm his fatherhood by giving gifts to the mother and the child. In many instances, particularly when a woman settles with a man, her children by other men may go to live with their fathers or their fathers' kin. Thus, although Caribbean societies may be said to have a fluid institution of marriage, the principle is not lacking, namely, that by various customary or formalized means, a pair bond is acknowledged so that the adult male can achieve his identity as father and the web of kinship can continue to be woven. As mentioned above, the Tsimihety of Madagascar frequently practice "trial" marriage, whereby a male and female cohabit for a year. If they prove suited to each other, and if a child is born, this cohabitation is confirmed by rituals and exchanges. But if the male does not become a father he can insist on a peremptory annulment.

The conditions of marriage, understood in the broad sense I have discussed above, simply complement the incest taboo. Whereas the taboo proscribes sexual relations between persons tied by the primary bond and by the original kinship relation of father, marriage prescribes sexual relations between persons who are not tied by either a primary bond or a kinship tie. The pair bond, as the precondition of kinship, is thereby kept separate and integral. But as the Caribbean example makes clear, marriage as a cultural ritual is transitional, intended to assist in the transformation of adult male into father (and of female into mother). In this sense, the incest taboo is also transitional in function, since it proscribes the fusion of identities established separately as complements in the primary bond.

The principles of the incest taboo and marriage therefore both work to ensure the continued existence of kinship. No such guarantee exists or can be extracted from the natural sexuality and nurturance behavior of the species, since the stability of sexual attraction between males and females is threatened by the continuous receptivity and readiness typical of the species. It will be appreciated, I hope, that I am not arguing for the "natural" origins of kinship and associated institutions. Empirical evidence suggests that kinship has been the first principle of human social life, but it is cultural: there is no warrant for maintaining that kinship must and always will remain in this position of primacy. True, if kinship has been primary from the beginning, it must be deeply ingrained in human culture and thinking. But although it is the fundamental adaptive principle, its effectiveness will last only as long as the environment remains constantly receptive. If, whether by natural processes or by human influence, the environments of human populations change to the extent that kinship ceases to be effective, then it may wither, and along with it the incest taboo and marriage.

To some extent such a decline has become apparent to those living in urban environments, for whom the pressure to find an alternative basis for social organization has become greater. Those who continue to live in a rural environment, whether in Euro-American societies or in those of the Third World, continue to find kinship adaptive. Whether the principle of kinship is questioned by a total society will depend on how far environments are changed so as to render kinship maladaptive in general. One can sense the problem in the various attempts to construct alternative life-styles (often modeled on the family) and in the increasing expansion of welfare agencies that negate "family." And since kinship is associated with the control of sexual relations, one can also sense the emergence of an environmental problem relating to kinship in the new technology of contraception. In a paradoxical sense contraception may be thought to have rendered the individual sex act more "cultural" while making human sexual activity more natural. Thus it has removed much of the authority of kinship over sexual relations. And if I may be permitted a final, marginal observation, it might be said that if women achieve cultural parity with men, so that culture is no longer dominated by the male idiom, then the integrity of the species (both sexes) may be preserved only through the exploitation of a technology that permits men to gain a biological parity with women by establishing primary bonds with offspring through nurturing. This has been possible since the advent of the formula and the bottle.

3 THE PROMISING PRIMATE: Content

Consciousness

The recent field studies of gorillas and chimpanzees by such workers as Schaller, Fossey, and Lawick-Goodall have been conducted on the model of anthropological fieldwork in alien communities. The observer seeks first to efface himself as a cultural influence and second to allow himself to be molded into the ways of the community. As a result we know more about animals in their natural state than we ever dreamed we could. One result of this intimate acquaintance is that these animals have become known to us as individuals; one might almost say that they have become humanized. They are known to us by name—Flo, Leakey, David Grey Beard—by personality, and by idiosyncratic things that they have been seen and recorded doing, whether sorrowing over the death of a close relative or raiding a tent for cardboard to chew on. But then, as everyone knows who has house pets, familiarity with the details of their existence reveals that each animal is somehow an individual. So even if we discount the more obvious ways in which nonhumans are transformed into beings with human characteristics (including naming and the implicit attribution, by the use of human models, of consciousness and intention to acts and emotions), it is fair to say that nonhuman primates, if not all animals, have individuality in appearance and in personality. These differences are apparent not only to human observers but also to the animals themselves, to the extent that patterns of association and interaction are founded on the attraction and repulsion of individuals.

Experiments in teaching symbolic forms of communication to nonhuman primates have shown that these animals have minds of their own which, given the means, they can express in "language." They prove perfectly capable of making choices and of showing preferences, particularly for certain foods, and also of expressing opinions (for example, as to whether something is bad or good, sad or funny). Such evidence bespeaks the development of what we might call ego, of a sense of judgment that enables the individual to differentiate himself deliberately from others and to recognize (though not necessarily approvingly) that he and they are different in some respects. This possibility now seems all the more likely, to judge from the experimental work of G. G. Gallup (1979), who has found that chimpanzees can learn to recognize their reflections in mirrors, while other apes cannot.

This evidence, together with more general indications of differentiation in hierarchies, gives the lie to claims that humans have sole monopoly on "consciousness," on being aware, on being able to regard themselves simultaneously as subject and object.

Yet if nonhuman primates, not to mention other animals, are individuated and have the power of individuation, how is it possible for one species, the human, to develop consciousness, and particularly self-consciousness, to such a degree that it becomes of critical importance for the individual's sanity and survival? And what is the meaning of this development in and for human evolution?

There are two matters that I wish to clarify separately. First, "individuation" means that individuals are seen to possess characteristics that make each one different. This perception is utilized as a basis for attraction, repulsion or association. Such individuation, considered in its totality, may be called variation, and it is the basis for selection of a mate. Differences that are perceived may or may not reflect others that are hidden or genotypic, and insofar as the perceived signals the presence of the genotypic, individuation is simply the surface and instrumental means of natural selection.

The second concept, self-consciousness is far deeper and subtle and may be equated with individuality or even personality. It is not entirely dissociated from individuation, but it does have a qualitative difference. For self-consciousness means that an individual perceives differences *in himself,* regarding himself as a complex and heterogeneous entity made up of separate but interrelated parts. A self-conscious individual is in a sense dissociated, in the first instance perceiving "himself" with another "self," able to catch one part of himself with another part and, in R. D. Laing's term, a divided self. The main division of the human person separates what he sees of himself, the surface of his body, from what he cannot see but supposes of himself—what is inside. The "inside" is most frequently the subject, that which perceives yet can also consider itself an object, something to be perceived or conceived. What is within is frequently acknowledged to be the owner of the body yet is also thought of as being possessed. We can speak of "my mind" or "my spirit" as easily as "my body." This self-consciousness or individuality is the basis of bonding and of permanence in relationships.

But how is self-consciousness possible? What evolutionary conditions in the constitution and environment of the early hominid came together, formulating a problematic that made such consciousness adaptive? We might choose to cite certain suggestions that language is the prerequisite, for it is only with the aid of language that we can find the way to give reality, by articulation, to the inchoate intuition of the divided self. But language may play this role only in a mechanical sense, by providing a means of *expressing* and *symbolizing* consciousness. The sense of self may still exist without symbols; the notion of language presupposes that there is something to articulate (I realize that some would maintain that we only know whereof we *can* speak, but I shall stand by my position). But even if we choose to account for self-consciousness by the advent of language, the question still remains: what makes language possible and adaptive in the original context? The accomplishments of language in succeeding contexts will not explain this.

In all probability, language and self-consciousness were, as they are now, bound up together, and maybe they evolved together. If so, an exploration of the origins of self-consciousness may reveal the origins of language. The discussion of the divided self proposes as one implication of self-consciousness that the individual can be regarded and can perceive himself as partible, and we might conclude that individuals so regarding each other and themselves can share themselves. This idea provides us with a useful point of entry into the evolution of self-consciousness, since we can now inquire into the conditions and problems that make such reciprocal sharing adaptive.

There is an added justification for this starting point. In spite of the evidence for individuation among nonhuman primates, we lack any indication that they *share* either themselves or physical objects. They lack any disposition to acknowledge an obligation to share, or if you like, any sense of reciprocity. There does exist evidence that nonhuman primate individuals will show solicitude toward other individuals and that they may admit forlorn strangers into their group and territory; in this sense they could be said to share. But there is no evidence that such "altruistic" acts reflect any sense of obligation or any expectation of return. Still, this reciprocity and mutuality, the essence of human sharing, is missing from the chimpanzee practice described by Teleki (1973). Chimpanzees may beg or even snatch meat from each other without reprisal. But there is nowhere a sign that he who gives on one occasion can expect on another occasion to receive from him who has taken. Chimpanzees in their natural state do not formulate a concept of the individual as divisible. For only with this notion can an individual think of himself as existing *in the future,* and only if he thinks that he *will* exist can reciprocity and mutuality with respect to objects become effective.

The notion of sharing and its ramifications have for a long time been seen as a key feature of human productive activities. Sharing is the basis of cooperation and competition, for it carries with it the idea of the division of labor, a process that makes possible a far more efficient means to material production than individual self-sufficient activity. Cooperation has also been identified as a foundation for the organization of human protection and social living as a whole. Having been recognized as functionally efficient human activities, sharing and cooperation have hitherto been introduced into discussions of human evolution as if they had simply come out of the blue, as some sort of miraculous event that accounts for human superiority. But if sharing is so fundamental and so human, a theory of human evolution must explain how such a practice is possible, and present theory does not do so. The usual explanation states that the development of hunting requires cooperation; but we may object that if it requires cooperation, then it presupposes cooperation. It is sometimes also argued that the practice of hunting is eventually found to be more efficient if individuals cooperate. But this does not apply to the hunting of small animals, nor is cooperation necessary if a spear or some other projectile is available. In any case, if an activity evolves through cooperation we cannot consider it the cause of cooperation, and we must return to our still unanswered question. Let us approach it by reexamining the characteristics of individuals and the ways in which they organize themselves socially.

Individuation

Nonhuman primates do not possess the qualities of sexuality; they do not manifest permanent sexual characteristics that serve to differentiate individuals and function to establish ties of mutual interest (though there are, of course, sexual dimorphism and different genitalia). Nonhuman primate sexual activity is periodic and, as we may recall, is particularly tied to the waxing and waning of estrus in the female. During estrus, parts of the female body assume sexual significance and become the focus of attention of males. But once this period is over, the mutual interest of males and females in each other for the purposes of sex (or more accurately, of reproduction) virtually ceases. Male and female live separate lives, united by bonds of proximity and protection. This is not to say that nonhuman primates do not recognize each other as individuals, but rather that this recognition is not based on permanent and idiosyncratic sexual characteristics.

In the human primate, sexual receptivity and interest are constant and long lasting. They are accompanied by, even signified by, the development of permanent sexual characteristics common to the species and varying in degree from individual to individual. The adaptive advantage of this generalization is quite clear—reproduction is possible in a wide range of environments. For example, any two mature individuals of the human species can mate in any climate or geographical zone quite independently of seasonal variation and influence.

The permanence and continuity of sexual receptivity and attraction elevates the sex act and sex differences to prominence. Sexuality becomes a defining and uniting feature of the species. Through it, males and females in general maintain an interest in each other, while individuals who display variations utilize this common interest and these variations as means of selection and of differentiation. Sexuality, then, is a primary source of individuation in the human species. No doubt this phenotypic variation can be linked with genotypic variation and can be explained as a mechanism for broadening the scope of selection, a cunning device of the gene. But what is important for evolution is that the surface differences associated with continuous sexuality are visible and make an impact on individuals, serving to signal individuality.

As we have already indicated, this generalized human sexuality, though adaptive, is not immediately or automatically so. Generalization makes possible adaptation to a wide range of environmental problematics, but it does not define the precise mode for any particular environment. The continued sexual receptivity of females and the continued possibility of attraction between the sexes do not themselves offer a workable procedure. For if all males are attracted to all females all the time, it is doubtful that any reproduction could take place, and we would probably have a Hobbesian war of all against all. Natural selection on the basis of sexuality takes place through the attraction of individuals to each other, according to their individual variations. To put it simply, among nonhuman primates an individual with a small mouth, large genitalia, and a crotchety disposition may be neither more nor less attractive to another individual. If it is a female in estrus, she will be mounted. But such differences do matter to human individuals, and natural selection operates through the mediation of individual selection.

Human males and females are both separated from each other and joined in mutual interest by their sexuality, unlike nonhuman primates. This human trait, to which the term "sexual generalization" refers, entails a degree of differentiation according to sexual characteristics. Thus a natural basis for individuation provides some of the conditions for the possibility of individualization and self-consciousness. Nonhuman primates do not seem to experience sexuality or sex drives: there does not appear to be an association between feelings of pleasure, emotions of satisfaction, and copulation. We have evidence of heightened feelings of aggression, particularly among males, and heightened sensitivity and edginess among females who are in estrus, but since sexuality is not a focus of individuals' interest in each other, the opposite emotions and their fulfillment do not seem to be linked with sex in nonhuman primates.

The extraordinary attention that sex receives in human populations, whether it be the subject of repression and shame, as among civilized Puritans, or of open discussion and planning, as among the "primitive" Mehinacu of the Amazon (see Gregor 1977), is surely evidence of the significance of sex for individual emotions, which has a species-wide application over and above the meaning of sex for reproduction. This aspect has some relevance to self-consciousness, since an individual must regard another as a source of satisfaction and regard himself at one stage as unfulfilled and at another stage as satisfied.

Human sexuality entails human self-consciousness, however, in a purely selfish sense: two individuals regard each other as sources of their satisfaction, recognizing that each person can treat him or herself as a divided being, as one who exists for himself and whose self has meaning to another. And to the extent that this is so, we can speak of self-consciousness as having roots in sexuality. But it is a limited extent, since the relationship is a dyadic one, and not even the fact that humans can mate plurally or can be attracted to several individuals at once—or even that they can gain purely physical sexual satisfaction without emotional involvement—makes possible a self-consciousness involving the essential, triadic social relationship. In other words, the human sexual act itself, for all its emotional entailments, can still to some extent be fulfilled for particular individuals by a simple act of anonymous intercourse, as in prostitution. Human sexuality, taken independently of any other consideration, can¹ allow for promiscuity. It is only if we consider sexuality in relation to re-production and nurturance that the contribution of sexuality to individuality can be placed in perspective.

On its own, therefore, sexuality makes possible a limited achievement of individualization and self-consciousness—limited in the sense that it is necessary only for one individual to recognize that he or she can "give" something of the self at the same time that the other gives. The sexual relation need only be dyadic and immediate and so cannot be the condition for a self-consciousness that divides the person temporally as well as existentially or makes possible the sense of human sharing and social life. Because human sexual relationships are exclusively dyadic, they die when one member leaves; they are immediate, requiring the continuing presence of the participants; they are trivial or trivializing because minutiae and nuance cannot be distinguished from functional essentials; they are individuating, as we have already noted; and they are founded above all on emotion and self-interest. Such relationships were discussed in detail by Simmel (1950:126) who noted that sociologically, dyads cannot be social relationships, hence cannot be at the root of society.

Human sexuality is many layered, and although its function on the surface is to ensure attraction and selection, the deeper meaning of human sexuality revolves around reproduction, in which I include not simply giving birth, but provisions for the infant's survival to maturity so that he or she may continue the reproduction of the species. In this regard human sexuality is the foundation of the pair bond, and if we wish to explore the possibilities of human self-consciousness, we must investigate the conditions of relations between pair and primary bonds.

Although the two bonds are to be found in nonhuman species, they are found separately and not conjointly. Thus no possibility exists in nonhuman primates for the creation of the "father" and hence of kinship or for a conflict of interest between the pair and primary bonds. Thus among gibbons, male and female together raise their young, then expel them at maturity and raise more. Chimpanzee adult males are not bonded to females, and so a relation between pair and primary bonds, let alone a conflict of interest, is not even remotely possible. Baboons, with their harem groups, assure the male of sexual access to females, without running into difficulties concerning the relations between females and their offspring. There is, of course, competition among males for access to females, but it turns on the existence of hierarchy rather than on the competing interests of the female for her child and for her mate.

It would seem that a conflict of interest between pair and primary bonds is possible in the human species alone. And the crucial element is the dependent position of the male, who only becomes a full person through fatherhood, which necessitates admission to the primary bond between female and offspring—a position dependent, as we have noted, on the female's willingness to share the primary bond with the male. The transformation of this voluntary act of will into an obligation may be seen in extremis not only in patrilineal societies, which classify children without fathers as bastards or illegitimate, but also in those societies that clearly define identity as something to be achieved, requiring a man (in particular) to pass through a series of transformations, beginning with initiation, which brings adulthood, leading to marriage, which provides the transitional identity of husband, and culminating in the birth of the child, which brings fatherhood. Some societies (including the Tsimihety of Madagascar) go further, maintaining that a man in particular is a complete person only when he becomes a grandfather.

The conflict of interest between pair and primary bonds appears in the asymmetry of male and female relations that is made explicit in many human societies. In cultures that punish females more severely than males for adultery, the deep reason is that a female's adultery undermines a given male's standing in relation to offspring and, in a general sense, offends against the social premises by which kinship and social organization are activated. We must recognize that although human sexuality provides a base for continuous sexual attraction between the sexes and so makes possible the formation of pair bonds, it does not of its own accord necessarily continue between two attracted individuals for their lifetimes, nor does it preclude the possibility of one individual's simultaneous or serial attraction to several others. And in this regard recent research suggests that the female is less easily fulfilled than the male and is therefore the more inclined to resist the privations of a culturally transformed pair bond. But although alternative and plural sexual relations may disparage individual egos, they would not offer reason enough for social sanctions, which result when the possibilities of kinship, hence social structure, are brought under attack. Let us therefore return to the conditions of the relationship between primary and pair bonds and its function in the realization of specific forms of social organization.

The male formally becomes a father only by reference to the female who has offspring. The corollary of this is that the female is obliged to share the bond with her offspring if the necessity of fatherhood is acknowledged. This situation might change, but it has yet to do so. The obligation of the female to admit the male means that she must defer to the male. Both the referential aspect of fatherhood and the deferential aspect of female status emerge from the existence of a third person, the offspring. (A male pair bonded to a female without offspring is not a father, and she has no primary bond to share.)

Thus if the possibilities inherent in the metarelation of pair and primary bonds are to be realized as fatherhood and thence kinship, the transformation of individuals depends on reference and deference. The transformation of male into father is real only insofar as it is known by the individual and recognized by other individuals. And they know by referring the self to another self, which in turn is defined by reference to a third self, the offspring. Similarly, if a female is to know herself as the person upon whom obligation falls, she can do so only through deference. And the same is true of offspring, whose identity as offspring is established with the mother, but not with the father. The father's identity is established only by reference to female and offspring, but their relation to the father is instituted through deference, a willingness or obligation to admit him. Thus the sense of possession in relation to self becomes apparent. The primary bond is unitary but can be shared without destroying the unity. A female shares her title to offspring with the male, and this detachment of offspring places them in his possession. For children in general do not make him a father, but *his* children, those whom he possesses and to whom he must refer if he is to know himself as father and be so known.

It is the situation of the female that leads us to the genesis of self-consciousness. The metarelation of primary and pair bonds, on which so much that is social depends, revolves around the female, who is the common member of both constituent relations. When she admits the male to her bond with offspring, she divides or shares her bond with the offspring. The female undergoing this process must come to see herself as divisible; within her, one part, "wife" or "mate," can offer a commentary on the other part, "mother," and vice versa. The pair bond is founded on sex, while the primary bond is founded on nurturance, so that behaviorally the female is compelled to a consciousness of her existence as one bound to others in quite different ways. In the performance of the different sorts of attention that characterize the pair and primary bonds, the female can very easily find herself in a contradictory situation. She must make a decision as to which function to perform at a given moment, without having that decision negate her total, double function as female. In the most primitive sense the realization of the pair bond through sexual attention (not necessarily intercourse) may be prevented or at least modified by the demands of offspring for nurturing attention. In this simple case, an order of priority has to be established without threatening the life of the relegated relationship. The continued existence of the metarelationship depends upon a working out of the performance of the constituent bonds relative to each other, and if one bond is realized at one time, some guarantee or promise must be given that the other bond will be realized at some future time. If a structural possibility is to become a functional reality, then individuals must offer each other promises of their future intent that will be received and reciprocated. To offer a promise is to presuppose cognizance of a self that exists now and will exist in the future, and the first evidence of this is in the situation of the female—it is the female who "gives" herself to the male and to her offspring, who divides herself between them through promises and their fulfillment.

Although the genesis of self-consciousness in this situation is primal, it does not exist formally because the human situation is one that emerges only as this primal situation is transformed. The human situation or the cultural situation is not the metarelation of primary and pair bond; these are the conditions of its possibility. The transformation into kinship marks the point at which the human emerges, as the emphasis shifts from the primary and pair bonds to the kinship bond of father/child. At this point the child replaces the female as the new joint for kinship relations. A series of transformations occurs: the male becomes "father" (a different order of connection from any existing before); the offspring of the female acquires a kin identity (child, son/daughter); and the female now bears to the offspring the kinship relationship of mother. The node of kinship reverses that of metarelation: the relations of kinship must pass not through the female, but through the child, who connects the bonds. In this way the child becomes a divided self, linked to father and mother, its primary kinsmen. Here again, division in the elementary situation brings the possibility of conflict between the double allegiance of the child to the mother (as kinswoman and the other half of the primary bond) and allegiance to the father (as primary kinsman). Since the category of father depends on that of child, yet is only a cultural, not a natural, category, the weight of obligation (or law) is on the side of the father. No matter what else happens, the father's title must be protected.

After the kinship relation "father" has been established, the male becomes the node of intersecting relations. He is at the same time the primal kinsman to the woman's offspring and the other party to the pair bond. Since kinship is possible only through the condition of the primary/pair bond metarelation, neither of these can be conflated, nor can they be merged with a kinship relation. If the pair bond were to be merged with a kinship relation, this would mean that kinship had existed prior to pair bonding, presupposing incest, which is nonsensical. Thus the male is also a divided self, having to see himself as at least two different "persons" relative to others to whom he is bound.

Let us pause briefly here to make explicit certain critical and methodological points. It seems to me that human self-consciousness is something that is "personal" to the human species, if only in the simple sense that other animals cannot have a consciousness of being human. Anything that is personal to a species cannot have originated or have any meaning in any way other than through self-reference, that is, the individuals making up the species must think about themselves and must have been in a problematic situation that made thinking about themselves productive and adaptive. For this reason there is no point in seeking the preconditions for self-consciousness beyond the species itself and its member individuals. In other words, we need not consider other species or even such material factors as tools or hunting.

Second, the prevailing explanation for self-consciousness views it as deriving from cooperation, and this in turn is attributed to hunting and the subsequent need to share the kill. Although hunting among contemporary peoples is frequently cooperative, it is, as I have noted above, by no means exclusively so. But there are more important questions we might raise about the power of the hunting activity to create conditions for culture. Hunting is almost universally a male activity, often metaphysically or mystically characterized as a male monopoly of death (in contrast to the female monopoly of birth). Although females assist in some societies (for example, the net-hunting pygmies or caribou-hunting Eskimo), the predominance of the male as a hunter seems unquestionable. If this is the case and if cooperation, language, and self-awareness are ascribed to the development of hunting, the female is relegated to a cultureless, passive role. There is no evidence whatsoever that women in hunting and gathering societies are any less cooperative, verbal, or self-conscious than men, nor that any of these qualities that they possess are learned, or merely rub off on them, from men. Evolution refers to a species, which includes both sexes and their generations.

Third, it has become increasingly evident that with the exception of extreme environments such as the Eskimo, there is no such thing as a hunting economy. Hunting is associated with gathering, which is primarily a female activity but from which men are by no means excluded. Although meat makes a definite (and preferred) contribution to the diet, it does not constitute the major part. So I cannot see the justification for arguing that an activity commonly involving only one sex and marginal to subsistence is the primary activity generating culture. I shall offer an alternative view of the significance of productive processes to human evolution below. Still, in the present context it is noteworthy that even if the proposals outlined above are incorrect, they represent an inquiry along the right lines: the realization of self-consciousness and culture from natural possibilities must be associated with that aspect of the human evolutionary problematic which demands that *the species consider itself part of the problem.*

The Promise

The primary and pair bonds are natural conditions, as much a given part of the human situation as upright posture or the opposable thumb. Their conjunction or overlap is a natural effect rather than a direct natural fact, and its existence therefore depends on other conditions. Still, however it originates, this conjunction inaugurates a promising situation, a possibility. Part of the species problem of generalization pertains to social organization and reproduction, and the metarelation promises to provide the means for development of a universal cultural principle of social organization sufficiently general and flexible to allow for the emergence of a variety of forms. This means is kinship. The evidence concerning the size and possible formation of the brain of *Homo erectus* supports the proposition that the creature was intelligent enough to transform the promise into a reality. I am not saying that kinship was there, but the promise was awaiting realization, and the metarelation had to be made into kinship. Given that the promise applied to the entire species, we need not suppose some original genius, nor need we suppose a long, drawn-out process of trial and error. Still, it is important to recognize that kinship did not emerge of its own accord but that early hominids created it out of natural raw materials and that this creation involves technique. The prerequisite is not manual dexterity such as is necessary for tools but a skill by which the individual can remake himself and others—again, not by deforming his body but by refashioning his idea of himself. And the final tool necessary in this process I shall call the promise.

Let us first consider this tool in its general context. Human kinship involves individuals in relationships defined according to distinct criteria, with each person as the junction of at least two relationships. Each individual sees himself as divided and as owing different forms of allegiance to others—a male is father to the child and spouse to the female; a child is primary bond partner to the female and kinsman to the male; a female is primary bond partner to the child and spouse to the male. Unless these forms of allegiance are kept distinct and separate, no one relationship can enjoy autonomy and privacy—no one relationship can fulfill its function—but will at some point come into conflict with the other relationships to which it is connected. To oversimplify, if the demands on the female as spouse never abate and the relationship itself never assumes a quiescent state at some time, then the relation mother/child cannot be given its due, and the child's identity (as well as its emotional condition) can only be partially developed. The same may be said of any relationship relative to the coexistence of the others. A mother who immerses herself in the primary bond cannot fulfill the pair bond, with resulting consequences for the other, associated identity, that of spouse. These are crude examples, perhaps, but they illustrate a general and real phenomenon in real social situations in which individuals suffer, as does the structure built on their liaisons.

Thus the elementary social situation is a plural one whose continuity and livelihood depends on the separation of parts and their sharing of attention. People enact each constituent relationship at certain times or under certain circumstances *on the understanding* that at other times it remains in the background, undefined.

When it is in the background, we accept this dormant state *on the understanding* that at a future time, or under certain conditions, the relationship's time will come, and it will assume the spotlight. Without this understanding, all the relations to which an individual is party will make equal and competitive demands, with the net result of a Hobbesian emotional war of all against all.

This separation of relationships is the public side of the individual's divided self—the individual knows the several aspects of his "self" only through his relation to others. But whereas the public demand is for separation, the individual must keep the facts of himself together as an integral whole, and to do this he must depend *on the understanding* others show in demanding of him only an appropriate self and not tyrannizing one self to the total exclusion of others. In turn this depends on the individual's willingness to share himself with people so that they may achieve their own sense of self, their own integration.

By approaching kinship and relations from this angle, we see that the condition for its existence is summed up in the phrase "on the understanding," for there is nothing in the natural condition of the individual that can provide the guarantee of the separation and integrity of relationships. There is no control mechanism comparable to the hormonally conditioned onset and cessation of estrus that will bring relations to the fore and replace them in the background. The understanding and the will of individuals alone can be utilized.

Equipped only with an understanding, and without an infallible guarantee, individuals who depend on their relations to others must at the same time accommodate themselves to separation, and separation from those to whom one is bound is a primary source of fear. This is particularly true of those who lack understanding—young children, for example, who are the partners of the primary bond and the elementary kinship relationship. This fear, as Bowlby defines it, is "an instinctive response to one of the naturally occurring clues to an increased risk of danger" (1973:86). It is also a principal source of anxiety, even though from an outsider's point of view, there is no imminent or real danger. Separation requires only that the attachment figure, the relative, be inaccessible physically, emotionally or existentially (though physically present, the partner's attention is turned elsewhere; see Bowlby 1973:23). When we remember that we are concerned here not only with the psychological well-being of individuals for their own sake, but with the structural well-being of the source of kinship and culture, we must add that the lack of integrity, separation, and firm connectedness of relations is a source of anxiety and fear for the "system" itself—any individual's action that might threaten the integrity of relations and their connectedness is a source of anxiety to outsiders, a matter of principle to them. So this understanding of which I have spoken has its location and being in the mind of the individual and must in addition be acquired by him as an *obligation,* clarified by the existence of sanctions, so that if the individual fails to exercise his understanding, this failure causes other people to impose their understandings, on his, and their, behalf.

The individual's understanding is that he or she will permit the partner in a relationship to disengage from that liaison in order to engage with another partner. This permission will be given only if there is a basis for confidence that at some future time the relationship in question will be renewed. Thus each individual must share himself with others or sacrifice something of his title to relations with others. The exercise of this understanding assumes, with some faith, that others will also exist in the future. The self is divided temporally and laterally, as it were—that is, across time and with respect to simultaneous occurrences. The partners who are disengaged will be set free only on condition that they understand their obligation to suspend another relation in order to reengage the relation from which they have been temporarily set free. They too must have some idea of themselves as divided laterally and temporally before this can be undertaken or the conditional situation can be regarded as possible.

The establishment of human social relationships and the assurance of their continued existence, no matter what form they take, depends upon the exercise and imposition of the promise. Thus we see how man may be regarded as a promising primate in two senses. All that is made possible by the realization of kinship in the human animal, the promise of society and culture as adaptive dimensions, rests upon the development of the ability of individuals of the species to commit themselves in this way.

Yet promises, being inventions and not a natural property of the human mind (Hume 1739/1888:519), though requisite for society, cannot have the force of a natural necessity, and if society cannot be sustained unless the promise is carried out, there must be another, independent form with the function of serving the promise by imposing the sanctions that will enforce it. In Hume's words, again, "We must feign a new act of the mind, which we call the willing of an obligation" (1739/1888:523). This sanction, the public side of the private promise, is the institution of the taboo.

Breaking promises would result in the self-destruction of the relations constituted by the promise. At the individual level this would be the withdrawal of trust—"When a man says he promises anything, he in effect expresses a resolution of performing it; and along with that by making use of this form of words, subjects himself to the penalty of never being trusted again in case of failure" (Hume 1739/1888:522). But the withdrawal of trust follows as a consequence of the failure of the promise and is, if generalized, the same thing as the disintegration of the structure of relations. Trust and its withdrawal are dependent on the promise, not independent of it, so although the withdrawal of trust may be the consequential penalty, it cannot be the autonomous sanction that can be exercised *before* the promise is made. If the failure of promises amounts to the disintegration of structure, then this failure must be regulated by a sort of built-in safety device that is part of the structure but will not disintegrate. The taboo must therefore have a more rigid and pronounced degree of institutionalization, because it must have an autonomy far in excess of that typical of or necessary for the promise. I think that this reasoning explains not only the existence of taboo but also its precision as an institution in comparison with the promise.

Irrespective of their particular form, taboos as prohibitions of an act, an object, or a person presuppose existence in a positive form. When a sexual taboo is applicable to persons in a given relationship, it presumes that the relationship already exists, as does the possibility that the condemned behavior can or will occur. Thus the taboo does not define phenomena but bases itself on a preexisting definition. The object is to preserve and guard against the dissolution of social facts by the failure of promises, to prevent the dissolution of the boundaries between social phenomena that will result in conflation and preclude continuity and perpetuation. In the primal instance this aim applies to the metarelations between primary and pair bond and the immediate kinship relation of father. The incest taboo "belongs" to the society already constituted by kinship yet is imposed autonomously, on behalf of kinship itself. We should understand in this sense Lévi-Strauss's assertion, which is also Tylor's, that the incest taboo compels exogamy (marriage or mating out of kinship). Exogamy is not the consequence or the effect of the incest taboo but rather a form of the promise, that understanding which is a matter of will and principle. The taboo exists to maintain the integrity of the promise and ultimately of those who make promises. Phrased slightly differently, people do not deliberately seek to avoid incest; they seek out nonkinsmen as sexual partners and thereby avoid incest. But incest always remains a possibility, and the taboo exists as a formal feature to negate, not the possible occurrence, but the possible occurrence *in public* in such a way that kinship itself, a community (and species) property, is threatened.

Since the integrity of relations is in part a function of the divided self, whose parts retain their apt reciprocal connections to others in relationships, the taboo also serves to preserve the coherence and integrity of the individual's identity both in its complexity and in its dependence on integration for emotional stability.

In the promise and the taboo I find at the abstract, conceptual level of human species existence the focus and junction of the individual and the social. It is not that serving the well-being of individuals coincidentally ensures the survival of the social, or that assuring the survival of the social serves the well-being of the individual. Rather, in the promise and the taboo the individual and the social merge while preserving their own integrity, doing so through relations in which the individual is identified as a divided self and through which the social system is constructed, when relations are related to each other.

Before we can conclude the inquiry into the genesis of social structure, we must ask one more question, even if we cannot answer it properly. How is it possible for the individual to know himself as divided, and how is it possible for related individuals to undertake promises and to impose taboos? In all these matters, as I have already noted, there exists some idea of the future as a prominent feature of the present, as something which has to be taken into present account. Further, if any idea of the future is involved as part of the present, then this must automatically assume a knowledge or recognition of the past, since the present is that coming into being of the future posited by the past. We have to assume that the early hominids who originated kinship (and tools), and who had the brains to do so, also possessed certain skills or techniques that they developed at the same time and most likely in relation to the development of other aspects of culture.

Technique: The Production of Reproduction

When we find early hominid tools, we presuppose that early hominids knew how to make them and had the skills to do so. It is part of the present-day archaeologist's research technique that he learns how to make stone tools, and it is not an easy task; there is a knack to napping (see Washburn and Moore, 1974). Yet when we think of inventions like kinship, group structures, promises, and taboos, we tend to forget that as social objects they have an ontology, they had to be brought into existence and so required a technique, determined, planned, and acknowledged as the means to an envisaged end.

There is a difference, of course, between a stone tool and a kinship identity, but this should not be allowed to obscure the fact that both have to be manufactured, repaired, and serviced. People must do something to someone to create kinship or, for that matter, any sense of identity. The crucial difference between the techniques of material and social engineering is that the latter is applied when human beings treat themselves (rather than nonhuman matter) as the raw ingredients and thereby transform themselves into "artifacts" (or identities). Of course, human beings cannot literally change each other beyond all recognition. On the other hand, it is not hard to change the appearance of a person by applying painted or tattooed designs, by adding decorations and ornaments, by scarifying, knocking out teeth, cutting hair, circumcising, or cutting off finger joints. Such practices we know from ethnography to be widespread among human populations and to be extraordinarily resilient. Ethnography also tells us that such practices are carried out under clearly defined circumstances and conditions that are subsumed under the heading of ritual. There exists in the majority of human societies known to anthropologists a technical activity with the object of changing the appearance of individuals so that they may be seen by others to have specific identities, while the altered individuals view the changes as transforming their identities. In societies that practice circumcision, for example, uncircumcised males are subadults who cannot marry or take an active role in the government of the community. They are also raw materials of the community, unable to function as tools of the social structure or of other individuals until they have been manufactured. We do not usually take this view, because human beings are involved, but certain things cannot be done unless there are enough male adults. The following example from the Ndembu culture provides a good illustration (see Turner 1967). Here young males who are not circumcised are regarded as effeminate or immature and cannot participate in running village affairs. If the proportion of females and uncircumcised males to adult males becomes too great, then the village, and especially its headman and the dominant lineage, is under threat. The major way of rectifying this is to initiate a Mukanda ritual in which young males are circumcised, that is, in which members of a community *produce* new members, just as if they had needed to produce more hoes. There are other possible means to recruit males to social positions; circumcision is certainly not the only way. But the point is that some action is necessary and any method used has a ritual character. Even in our own society, the establishment of the voting age by law is accompanied by a ritual action that people feel to some degree compelled to perform—a coming-out party, the presentation of a gift, and so forth.

The creation of identities and relationships is taken from the realm of possibility into that of actuality by a particular technique in which human beings act on human beings to transform them either from raw material into an identity or from one identity into another. This technique I shall call ritual, and it follows that until a successful form was developed, there was no successful social life. Furthermore, I suggest that the evidence we have of hominid tool-making activity, as early as *Homo habilis,* allows us to propose that such a creature was perfectly capable of engaging in ritual activity. Existentially there is every reason to suggest that the imperative to develop ritual, to establish the divided self and the obligations to share relationships, was more pressing and immediate than the need to produce tools. But until the late Paleolithic, we have no material evidence of ritual. Only the evidence of Neanderthal burial and cave paintings provides us with indirect evidence, in the sense that decoration and ornamentation we know to be among the accredited techniques of ritual.

By ritual, people express the idea that they can change each other and that each individual recognizes a change in himself. In the most obvious cases this is done by literally changing or deforming the body. But such a process of transformation, as we have thus far suggested, produces only what we might call an inert structure or situation. Individuals change from adolescent to adult or from bachelor to husband to father or from commoner to chief or from stranger to in-law. But unless there is some specification of what is to be done, such identities have no movement, no life. As we have seen, the elementary kinship situation, or indeed the establishment of any structure founded on relationships, works and serves the function of social living only when what we have summarized as the "promise" and the "taboo" is established. And neither promises nor taboos serve to define phenomena or to transform shapeless raw material into shaped and functioning tools. Both promises and taboos have at least one thing in common—they "make" something out of nothing. In the case of the promise, "that which may be in the future" is introduced as an object of the present. In the case of taboos, "that which is not" is made the object of attention.

I have already discussed the function of these institutions, and long ago Hume drew attention to the mysteriousness of the promise when he wrote, " 'Tis one of the most mysterious and incomprehensible operations that can possibly be imagined, and may even be compared to transubstantiation or holy orders, where a certain form of words, along with a certain intention, changes entirely the nature of an external object, and even of a human creature" (1739/1888:524). Although Hume was writing somewhat tongue-in-cheek, he expressed the idea crucial to our discussion. Now that we understand the mysterious nature of promises and their function, the question is: how are promises and taboos possible? Hume's answer is: in a certain form of words, which I will rephrase more generally: in some form of symboling.

Not until we meet with promises and taboos do we encounter the necessity for the symbol. I mean this not in a historical sense, but in a logical sense: the symbol does not originate until we arrive at the level of kinship structure or ontology signaled by the promise and the taboo. It comes at that point in the entire creative

process where the divided self and the fragments of selves connected as identities in relationships must include an understanding of themselves as existing in the future but at a moment in time
that is now. It arrives when the imposition of obligation is necessary to sustain the structure that has emerged from possibility and
when this sustenance gives the symbol life. In the promise, that
which is not but will be has to be transformed into the "now." But
I cannot promise to do something by doing it now. I must therefore give some indication in the present that is independent of the
future action but can stand for whatever I promise. I must symbol.
Similarly, I cannot establish what must not be by doing what must
not be done: I cannot impose the incest taboo by committing
incest and then negating it. How am I to negate it? I must somehow indicate what must not be by showing it not literally but
figuratively, by symbols and myths.

It is by no means accidental that we address the question "How
are promises and taboos possible?" immediately after the consideration of ritual as the technique for the creation of the social. For
the heart of ritual technique is the creation of abstract identities
by bodily transformation (or deformation), which allows people
to attain concepts of social phenomena by manipulating their
perceptions. At this level of ritual, however, there is still a material, positive aspect—by means of deformation, separation, and
inclusion, people are visibly and publicly made different. But with
the promise and the taboo, it is no longer the person who is the
raw material, but that which lies between persons, that which
must be taken from within the individual (his intent and will) and
placed between him and another, whose will and intent have
likewise been placed in the middle.

The actions of ritual are, to use J. L. Austin's (1976) term,
"performative"—as acts, they are ends rather than means; they
are what they do. When a man is scarified in a ritual context or
with ritual intent, his status is changed. That he may also be made
more attractive is coincidental. But though ritual actions are ends
in themselves, as actions outside the ritual context—dancing or

making music, for example—they may be performed to serve other purposes. The performative nature of rituals places them on the verge of literalness and utility: they could easily become symbolic, but they do not quite. They remain as they are, and they do not stand for what they are not. The same cannot be said of whatever institutes the promise and the taboo, for the promise and the taboo must assume an alien, symbolic (rather than actual) form. Something from outside their nature must come to *represent* that nature and content.

It is in the face of this aspect of the problematic, I suggest, that language, *the* form of symboling, arises, and not because promises and taboos cannot be indicated by some form of symbol other than the word (they can). Rather, symboling as a human activity derives its raison d'être in this area of the problematic. Since this is a subjective area, in which human beings do things to themselves rather than to inert or nonhuman materials, it is both economical and sensible to suppose that the human voice and the anatomy producing its infinite range of variation (its generalized property) may be exploited for the advantage of human beings in their efforts to adapt to the total, species-wide problematic. It seems equally sensible to propose that the emergence of language is contemporary with the discovery of other modes of signaling and symboling, that language should be considered as part of the total development of human culture, as, for example, the cognitive and motor skills necessary for tool making contribute to, and receive contributions from, the ritual skills of early human beings. Nor is the implied association, discussed above, between kinship, language, and ritual fortuitous. For here are three domains of human activity that share a single principle—they are generalized. They are never found empirically as kinship, language, or ritual, but always as kinship systems, languages, and rituals. They are supremely adaptive principles that allow human populations in widely varied environments to formulate specific solutions to specific problems, yet in the same general, that is, human, way.

I am all too aware that any attempt to formulate a theory of the

origin of language is doomed to failure. It has long been acknowledged that the invention and use of language is the single most
powerful contribution to human culture. It is by almost universal
consent the aspect that differentiates the human from the animal.
The many theories of the origins of language have given rise to
equally many criticisms. It is therefore pointless to embark on
extended criticism here. But the present argument confronts
some of the problems glossed over by most theories, particularly
the functional and utilitarian arguments based on the present-day
advantages of language. Current linguistic practice simply cannot
be taken as evidence regarding origin, for the simple reason that
the capabilities of language would have had to be known in advance. Language in its beginnings has to meet the requirements of
a problem existing outside language itself. The hunting hypothesis seems quite superfluous, since silence is necessary for
hunting, and therefore limited signs would suffice to engage the
cooperation of individuals. The ontogenetic theory holds that
language begins in the naming of objects. But while it is true that
human infants nowadays begin to speak by naming, they are surrounded from birth with names, objects, and persons. The ontogenetic problem of language is not the same as the phylogenetic, and we know that nonhuman primates manage to convey
information about objects over a distance without humanlike
communication.

Phylogenetically, objects of the environment are able to speak
for themselves; they manifest information that is accessible to all.
It is true, as Quine (1960) argues, that recipients of this information have no means of knowing whether they receive what others
receive, so they must find some way of exchanging their sensory
perceptions. But is this an original problem, associated with
survival—and therefore, by definition, a problem of the future in
the here and now? What can be seen, heard, touched, tasted, or
smelled does not pose difficulties, but what cannot be so
sensed—what is hidden, what is within, what is beyond—must be
rendered sensible, since it is not so by nature. Problems of death

and distance fall into this category and may have meaning to all individuals. Yet in the phylogenetic sense, and in the setting of the late Pleistocene, death must surely wait: living and continuing to live together, rather than as individuals, must be the more pressing problems.

4 DIETARY CONSIDERATIONS

Thinking about Objects

My argument thus far has focused on how the development of what might be called "subjective reasoning" was possible in one genus of primate, *Homo.* The very conditions of morphological generalization that permit maximum adaptation to a large variety of the earth's physical environments also require human primates to think about themselves. They must evolve, test, and execute means of adapting to the particulars of the environment. The same capacity that makes possible and necessary the development of varying adaptive strategies would not find it difficult to accommodate change, though some evidence, particularly that of tools, suggests that the earliest cultural adaptations were themselves sufficiently general to work for several hundred thousand years all over the world.

I believe that the major subjective problem for the genus, and particularly for the species *Homo sapiens*, was to organize reproduction and hence human social life. This is so because evolution classically emphasizes the survival of the fittest. The human problem, in the beginning, was ensuring not so much physical reproduction as the survival of those born by effective nurturance, and this difficulty was surmounted, I believe, by the development of kinship. The notable feature of human kinship is that although it may make some reference to genetic procreation, it is not bound by the logic of the genes. It is a flexible notion, able to organize not simply procreation but also numerous individuals, creating relationships that provide a basis for the planning of other survival activities. It is at one and the same time a mode of organization for procreation, food getting, social group living, including protection, and emotional satisfaction. Most important, it provides a single principle for including all generations and both sexes in productive and reproductive activities and for organizing them unitarily in all social aspects; it is highly economical, supremely flexible, and hence extremely adaptive. And although kinship in its entirety is a human phenomenon, it is not (as we have seen) a miraculous presence without traces among other primates.

Since a written argument demands the presentation of theses in sequence, that which I discuss first must seem to have priority. And to a great extent I would assert that the development of subjective reasoning has priority over the development of objective reasoning—but not necessarily a temporal precedence. Human reasoning, whether it concerns human beings or the nonhuman objects of the environment, is conducted by the same organ, the brain, and the same instrument, the mind. The fossil evidence of the enlargement and development of the brain from *Homo habilis* to *Homo erectus* to *Homo sapiens* is all the physical evidence we need to confirm that reasoning was a developing human trait. Next we must show what this reasoning was about, what facts or objects of the nonhuman, physical environments of the genus *Homo* were the most problematic and required, as a matter of course, sustained and genus-wide contemplation.

It should be noted that this last statement assumes the already existing ability to contemplate, to think, to solve problems. I shall be arguing not for the origins of rational thought but for its development to something we might consider a human level. As I shall show in a moment, there is every reason to think that other primates are capable of objective rational thought and that in fact they do practice such reasoning. But their environmental circumstances, possibly coupled with their genetically determined dispositions, have not made it hitherto possible or necessary for this reasoning to have developed as far as, or along the lines of, human objective reasoning.

I have no doubt that the primary reason is that nonhuman primates are more specialized than humans, hence more specifically adapted to more clearly defined and limited environments. As long as these environments do not change radically, adaptive patterns are not questioned, and primates have been given little or no cause to change their successful ways. But what is most interesting is that in the present century the environments of some individual nonhuman primates have been drastically altered, and these individuals have shown a remarkable capacity to change and to develop their propensity for objective reasoning. I refer in particular to the capture and subsequent imprisonment of nonhuman primates by human beings in zoos and especially in laboratories. These primates—not to mention rats, mice, and pigeons—have been very quick to learn to solve problems presented to them by a threatening species and in so doing come to strike us as being exceedingly intelligent, which is to say humanlike. For when scientists study animals, they set out to ascertain how close these animals can come to showing humanlike skills. Animals are rarely if ever compared and contrasted in terms of their own naturally adaptive abilities, on their own levels. Chimpanzees are not judged on their relative abilities at nest building or termite winkling, for example, but on how good they are at doing things that human beings do—better.

To understand the full import for evolution of findings concerning nonhuman primates in human captivity, we must understand that our presence in a zoo, laboratory, or field station radically alters their environment and presents them with problems that immediately touch their survival. They, in turn, respond with tactics that must be calibrated to the problems in this changed environment. Since the environment is humanly modified, these problems are basically associated with rationality, searching for efficient means to achieve desired ends. Thus animals from rats to chimpanzees show clear signs of developing their extant competence for rational, problem-solving thought.

Now, if we examine the new environments of these captive animals, we can see that the common and crucial feature is a drastic (often complete) modification of food availability. Not only are different foods made available, but the conditions for procuring food are also quite changed. No captive chimp need spend all day foraging for leaves and fruits but may be fed a meal at a certain time. Furthermore, much of the animal's training (teaching him to reason or solve puzzles) is achieved by presenting or withholding food. The heart of the problem of survival for such creatures is, therefore, how to procure food from this altered environment, how to adapt to new foods, to differing quantities, to altered timetables and to new ways of handling food. And in this new, problematic situation, the general solution is, indirectly, to cotton on to the moves of the invisible hand of the scientist and to develop reasoning powers and reactions that resemble human ways of thinking. If we think in terms of evolution in toto, the capture and imprisonment of nonhuman primates by human beings is an environmental catastrophe for those animals, a natural (rather than artificial) but telescoped change. If in the majority of cases we can elicit humanlike problem-solving thought from nonhuman primates by creating an environmental focus on food, then we might argue that these catastrophic changes could simulate the problematic from which early hominids developed their (hence our) powers of objective reasoning. The difference is that in the case of early hominids, the changes were natural (as are humanly caused catastrophes for nonhuman primates) and were not sudden, occurring over hundreds of millennia in widely separated and varied settings. The hypothesis is that human objective rationality developed in response to environmental problematics relating to food—its variety, availability, competition for it, and the relationship between such factors and the dietary competence of the genus *Homo*. To state the obvious, it is a question of food for thought, of the exchange of one for the other.

I begin the investigation of this notion by examining the dietary competence of those primates closely related to human beings, the apes. If our general argument holds that the primate order as a whole, and the anthropoids and hominids in particular, tends toward increasing generalization, then we should expect to find a more or less continuous progression of eclecticism in the diet of hominoids, with the widest range in *Homo sapiens*.

Hominoid Diet

The Gorilla. The gorilla can live in a variety of ecological niches within a narrow geographical range and has been recorded as eating more than one hundred species of plants, although the bulk of the diet is provided by one or two arboreal species. Schaller describes them as leisurely and fussy feeders, selecting only certain parts of plants and choosing from a wide variety available at any given time. Gorilla groups in different areas have varied diets. One group, living in the isolated Kayonza forest, was seen not to eat five plants that were eaten in other areas (Schaller 1965:357). This suggests that gorillas may *learn* their preferences. Since adults do not feed their young after weaning, they must fend for themselves from about the age of three (Schaller and Emlen 1966:377).

The Orangutan. "Although primarily a fruit eater [the orangutan] takes a wide variety of foods including leaves, young shoots, flowers, epiphytes, lianas, woodpith and bark. Occasionally animals eat mineral rich soils, insects and possibly small vertebrates and birds' eggs" (Mackinnon 1974:25). Mackinnon lists almost one hundred species of plant eaten by the orangutan in Borneo and a smaller number in Sumatra (1974:25–27). Orangutans show clear preferences for certain fruits and are very skilled and efficient at finding ripe fruit in spite of the unpredictable nature of fruit availability. (1974:31). Borneo animals, spend an average of 4.3 hours a day feeding, and Sumatran animals average

6.3 hours, a difference traceable to ecological conditions (1974:21). Infants cadge solid food from mothers, and the transition to solids is gradual (1974:46).

The Gibbon. This, the most widespread of the apes, is eighty percent frugivorous and twenty percent herbivorous, taking its food from the trees. The wide distribution implies an ability to select from a wide variety of possibilities. Gibbons eat for two hours or more and digest during three hours of rest in between (Schaller 1965:475–76, A. Jolly 1972). It also seems likely that in any one area the bulk of their diet consists of one or two staples or that in-season staples are selected at different times of the year. Insects, birds' eggs, and possibly birds are eaten as supplements.

The Baboon. Baboons are of considerable interest because although they are monkeys, not apes, they probably were forest-dwelling animals that later moved to open country—a displacement analogous to that probably made by the early hominids (see Pilbeam 1972). They have an eclectic diet, in keeping with their wide dispersal throughout Africa (Altmann and Altmann 1970:144). The baboons studied by DeVore and Washburn (1966) in Nairobi National Park and by the Altmanns in Amboseli subsisted on a diet of which grasses made ninety percent by bulk. In contrast, the hamadryas baboons of Ethiopia, studied by Kummer, fed for the most part on grass seed, flowers, and seeds of the acacia tree, ignoring the lush greenery of the nearby forest (Kummer 1968). Baboons of the South Africa Cape and of Amboseli were recorded as eating or sampling ninety-four species of plants (Altmann and Altmann 1970:148–49). Flowers, buds, young leaves, resin, and rotten wood (which may have contained eggs, larvae, or caterpillars) are the parts eaten. Though predominantly vegetarian, baboons have been seen to eat insects, small rodents, hare, and newborn gazelle; coastal animals eat a variety of shellfish. When locusts swarm, they are eaten in large numbers (Kummer 1968:168), and DeVore and Hall write that when army worm caterpillars infested Nairobi National Park, the baboons

ate nothing else for ten days (1965:46). Baboon diet, then, resembles that of other apes, being vegetarian and depending on
one or two staples supplemented from a wide variety of plants
that they are capable of eating. Meat is also eaten quite frequently,
though not regularly, being taken more or less as the opportunity
presents itself (DeVore and Washburn 1966:363).

Hamadryas baboons spend between five and seven hours a day
foraging and feeding (Kummer 1968:163–64), while grass-eating
baboons spend longer on the food quest in the dry season than in
the wet (DeVore and Washburn 1966:356).

It is particularly interesting to note that baboons show signs of
individual preferences for different foods. The Altmanns recorded: "On July 30th a mature female ate flower buds of a vine
like plant growing in depressions near *Solanum incanum.* . . . as
she ate she was watched by other baboons of the main group"
(1970:149). At other times individuals were seen consuming
eggs, lizards, chicks, and insects while other troop members
stood by.

Individual and group variations are important signs of dietary
flexibility among animals, which fits them to take fuller advantage
of environmental diversity and change. They indicate that good
choices may be learned, within an inherited limit, and the individual expression of taste suggests the presence of nondeter-
mined selection, an ability to discriminate voluntarily between
foods and between food and nonfood objects in the environment.
The baboons' broad diet and excursions into meat eating suggest
that primates can break through such restrictions as dependence
on flora. It would be possible, in other words, for the baboon to
avail himself of a range of fauna as well as flora to eat—if the need
arose.

The Chimpanzee. Chimpanzees are of the greatest interest to
us because they are genetically and behaviorally the closest to
man of all nonhuman primates (see Pilbeam 1972, Washburn and
Moore 1974). Chimpanzees and human beings bear a distinct
family resemblance.

Chimpanzees are eclectic in their diet. Reynolds and Reynolds recorded thirty-five plant species used regularly by animals in Budongo (1965:378–79), and Goodall, working in the Gombe, mentions more than ninety plant species utilized. Fruit makes up ninety percent in bulk. Feeding is highly selective; only ripe fruit or the tenderest buds are taken. Leaves, buds, blossoms, seeds, bark, pith, resin, and rotten wood may all be eaten, in addition (Lawick-Goodall 1974:267). Fruits ripening at different times of year or in different years or places are consumed in turn (Reynolds and Reynolds 1965:380–81, A. Jolly 1972:60).

The chimpanzee frequently eats insects including larvae, ants, and most famously, termites, for which it digs with a specially selected tool (Lawick-Goodall 1974). Bees' nests are also raided for honey.

Goodall describes chimpanzees as efficient hunters and writes of a group of forty individuals that they may catch more than twenty different prey animals in a year—mostly bush pig, bush buck, baboon, and colobus monkey. Though striking, this feat would hardly suggest that meat is a significant part of the chimpanzee diet or that hunting is a major feature of their repertoire. Thus chimpanzees *can* and *do* eat meat, they are *competent* to do so, but this is not a major *performance* in their dietary activities. It does seem, however, that if circumstances should prove conducive, meat eating and predation could assume greater significance in chimpanzee life.

So it is apparent that among the primates, the apes have a varied diet and select from a wide range of possible foods. The larger part consists of one or two favored and nutritious staples, as with the human dietary pattern. With the exception of the gorilla, all the apes mentioned have been seen to eat insects and so have a competence that extends beyond vegetable matter. In the case of the baboon and the chimpanzee, selection of fauna as food extends deeper and *under certain circumstances* strikes observers as being of significance for understanding human evolution. For it is the incorporation of meat into early hominid diet, via the development of hunting, that is so widely regarded as the springboard for culture. I will therefore consider chimpanzee meat eating in more detail.

Wrangham (1974) has carefully documented the occurrence of artificial feeding in the Gombe Stream Reserve, where most of the carnivorous behavior of chimpanzees has been observed. In order to habituate animals to the human presence and to facilitate study, Goodall and others provided bananas that were greatly appreciated by both baboons and chimpanzees. Wrangham notes, "During the periods of maximum feeding in Gombe National Park there was a very high frequency of attacks between chimpanzees and between chimpanzees and baboons" (1974:83). Teleki describes chimpanzee meat eating, sharing, and predation, relating them to the evolution of human hunting and suggesting, "The more we learn about primate behaviour, the smaller the differences between human and non human primates appear to be" (1973:42). In contrast, C. Jolly (1970) and Reynolds (1975) both maintain that the more we learn of chimpanzee hunting, the greater the problems for understanding human hunting. If hunting is a "natural" activity, how are we to explain the cultural development of human hunting? Teleki offers a simplistic one-to-one argument—chimpanzees hunt, ergo human hunting is a natural, programmed activity. And his reasoning is weakened by Wrangham's point, that chimpanzee hunting intensified in an artificial feeding situation. But none of the writers mentioned has seen the crucial implication of artificial feeding for human hunting.

Evolution of the Human Diet

We have seen that the diet of apes not only ranges widely among plant foods but breaks out, tentatively, from flora into fauna. The corollary is that such primates are competent to adapt to diverse environments and to accommodate changes in the availability of foods, coping with some scarcity, even of staples; these animals will have populations with a diet that varies from place to place. The generalized human digestive competence is simply an extension of a basic primate characteristic. We are omnivorous, but like other primates we retain preferences for some foods, and we select some rather than others as staples. Still, the exercise of preferences and the selection of staples are actions that betray the mechanism of *choice*. Apes are acknowledged to be fussy feeders, and we will shortly consider the clear evidence that chimpanzees have the capacity for making deliberate choices in the matter of food. Thus all the cognitive, rationalizing factors involved in preference, selection, and choice are not unique to the human primate; evolutionary conditions have made possible, and maybe necessary, the *development* of those factors. The most startling and important of these conditions are those associated with hunting, and with this in mind I will turn to an examination of the conditions that made chimpanzee hunting possible. This evidence may provide us with a model for the development of human hunting, and thence the development of objective rationality, the deliberate imposition of selected strategies to secure intentioned ends.

If we look at the situation in the Gombe Reserve from the chimpanzee's point of view, the introduction of artificial feeding by human primates constitutes, for the resident population, a dramatic but benign environmental change. Nor is it artificial: another species appears in and modifies the environment. As far as the chimpanzees (and baboons) are concerned, the environment has suddenly changed, presenting them with a highly delectable item of food in great abundance. But—and here is the catch—the environment is hedged about with restrictions that must be overcome if the available food is to be consumed. To get the bananas the animals have to *do* something. Bananas are favored by both chimpanzees and baboons. Whereas previously baboons and chimpanzees fed sympatrically, the one eating grasses, the other fruits, with the human introduction of bananas (not banana trees), the two species are brought into direct competition causing increased aggression and predation between them.

Given the tremendous overlap of diet among the primates, the interesting fact here is the development of competition for a common and preferred food item. It is fair to say that for a while, bananas became the staple for chimpanzees. And the possible analogue here for prehistoric human beings is that the development of hunting and predation may have been set in motion by their competition with others for a preferred or staple item, one that existed in abundance but over which one species sought to establish a monopoly. I shall take up this suggestion later.

With plenty of bananas introduced, but with their distribution controlled by human primates, Gombe chimpanzees have been presented with a tantalizing problem, one that arouses their desires and at the same time frustrates them. Goodall vividly and amusingly describes all the shenanigans in her research camp as the chimpanzees tried all sorts of ways to get the fruit. They were able to break into the storage boxes, prising open the lids with improvised levers, and they rummaged through the tents, turning everything upside down. They even fished into clothing with sticks. To keep the situation under control (by the scientists, that is) ways had to be devised to make sure that the chimpanzees did not have all the bananas they wanted whenever they wanted. In a film of the camp, chimpanzees are shown, staggering under the weight of enormous loads of bananas, dropping them along the way as they shuffle toward a sanctuary where they may gorge themselves without interruption. Initially, the chimpanzees were very successful, but their triumph was the scientists' failure, for the aim was to make friends with the chimpanzees by operant conditioning using bananas as rewards for services rendered. So the human primates sought ways to reinstate their control of the situation. I will give a telescoped summary of the strategies and counterstrategies that developed.

The first restriction on the bananas was to place them in boxes, but Fifi, Figan and, to a lesser extent, Evered

had discovered that, in order to open a banana box, all they had to do was to pull out the simple pin that served to keep the lever closed.

The humans sought to counter this:

> laboriously Hassan [an assistant] had worked at the fasten-
> ing, cutting threads into both pin and handle so that it had to
> be unscrewed rather than simply pulled out.

This worked for a couple of months,

> but eventually the same three youngsters had solved that
> problem too.
> Then Hassan had fixed nuts on to the ends of the screws so
> that those had to be removed before the screws could be
> unscrewed.
> Just before Hugo and I got back, Figan, Fifi and Evered
> had mastered that. Things were chaotic. [Lawick-Goodall
> 1974:143–44]

The humans then installed a large number of steel boxes specially made for them in Nairobi. These were battery operated and
could be opened by pressing buttons inside the research building
(1974:145).

This tactic was successful for a while, but the boxes had to be
filled. The chimps and baboons were by now accustomed to hang-
ing about the camp, waiting for feeding time, and they knew that
the boxes contained bananas. Aggression between baboons and
chimps, among chimps, and between chimps and humans increased:

> Aggressive interactions multiplied and, when the boxes were
> finally opened, there was bedlam. Something had to be done
> and done quickly. [Ibid., p. 146]

At first the humans stopped feeding the animals altogether, and
things quieted down. Then feeding was resumed on an irregular
basis and when it was judged that the baboons were sleeping far
away. The final solution is described by Jane Goodall in the following words:

Meanwhile work was going on for the construction of an underground bunker, stretching some ten yards from the main observation building. The finished bunker is about four feet wide and high enough for a person to walk upright. There is plenty of room for storing bananas inside, and boxes dug up from the slopes have been placed along either side. At long last we have complete control of when and whom we feed. [Ibid.]

What Goodall describes as shenanigans, and the trials and tribulations of primate research, do not seem so at all to chimpanzees (or baboons). Here was a competitive problem-solving situation between human primates and chimpanzees, over the control of food and its availability—and indeed over territory, for large numbers of animals hung about the camp and took the slightest opportunity to invade it. The chimpanzee purpose was to procure bananas; the human purpose was to gain control of chimpanzee presence in order to permit observation; and so humans had to find some way of controlling the salient feature of the environment, the distribution of a preferred food. The construction of boxes by humans and their opening by chimpanzees were *trial and error, challenge and response attempts at efficient problem solving by both sides.* Eventually, and not surprisingly, the humans proved the more intelligent and ingenious, but only to a degree. There is clearly a resemblance between humans and chimpanzees in practical reasoning, and it is not an accident that Goodall frequently observes how intelligent individual chimpanzees seem (by which she means how humanlike they appear).

There is yet more to this incident. Three chimpanzees made it their speciality to open the banana boxes, but one of these, Evered, frequently failed to reap the fruits of his labors, while the other two, Fifi and Figan, were more successful.

Evered would go up to a handle, unscrew it, and then with loud food barks, hurry to the box which he had opened. So of course, did any other chimps in the vicinity, and it was rare for Evered to get more than a couple of bananas, if that, unless he happened to be the only chimp, or the highest ranking one, in the camp. [Ibid., p. 144]

Fifi and Figan were more successful because they developed a strategy to deceive other chimpanzees. They soon realized that higher ranking chimpanzees would take advantage of their efforts, so

they just lay around, together with Flo, waiting for others to go. Then, when there were no adult males in the camp, they would quickly open a box each. Sometimes they could not resist going up to a handle and unscrewing the screw. But they did not then release the lever and hurry to open the box; they just sat, with one foot keeping the lever closed, casually grooming themselves and looking anywhere except at the box. Once I timed Figan sitting thus for over half an hour. But, of course, though the other chimps had not mastered the secret of the screws, they were bright enough to realize that, if they hung around, Fifi or Figan might, eventually, provide them with bananas. So they stayed in camp longer and longer. [Ibid.]

Here we see a developing competence for problem solving not only of a mechanical sort but also, among some individuals, of a tactical sort as well. Several chimpanzees (Figan, Fifi, Evered, and Flo) develop rational tactics that are aimed at maximizing self-interest by preventing forced sharing, and others cultivate a response based on watching and waiting. Shades of Hobbes! But also extraordinary is the point to which I shall return later, that the watching chimpanzees failed to learn how to open boxes for themselves—observer's knowledge was not sufficient, and those that had the skill made no apparent effort to teach others the secret. Of course, there is no telling what might have happened if chimpanzees had had a longer time to grapple with this environmental problem.

We may conjecture that the chimpanzee views the Gombe situation described by Jane Goodall in the following way. A foreign species immigrates and settles, and a preferred food item is made abundantly available but difficult to procure: there is competition from a formerly sympatric species, the baboon, for a preferred resource; competition from a foreign species, the human, which exercises and strives to maintain control over the resource; and an objective difficulty associated with procurement of the resource, in the mechanical nature of the container (a restriction similar to that presented by the natural containers of foods such as nuts). All these restrictions must be overcome if the ends are to be attained.

This environmental change is not a disaster but rather a benign catastrophe. To the chimpanzee the Gombe might be said to have become a veritable Garden of Eden. Its transformation coincides with an innate preference for certain foods, which in its turn governs selection and choice, but reinforces these dispositions by making them problematic. When the conditions of the environment are changed, we are able to see more clearly how food preferences motivate intelligence by demanding the application of thought and ingenuity as a guide for action to obtain ends, to realize preferences. At the same time, when the realization of these preferences requires competition with others of the same and different species, we observe a development of aggression and predation. Chimpanzees became more aggressive toward one another ("adult males were becoming increasingly aggressive"; Goodall 1974:143) and toward humans, and they became predatory toward baboons. It was under these sorts of condition that the deliberate hunting and eating of young baboons was observed, with signs that the chimpanzee competence for eating meat might have started to develop. In this respect it is noteworthy that there is no evidence that chimpanzees share meat (or bananas), though individuals may beg meat from one another and receive it. Although the cornering of a prey may be a cooperative effort, its consumption is an individual affair.

Nor is there any evidence, from this admittedly limited situation, that meat rather than fruit was, or showed signs of becoming, a preferred item. It was eaten as a supplement, and that the chimpanzees needed the protein is doubtful. The business of hunting among chimpanzees seems to develop more from stress than from necessity, but the stress may well be allayed in the course of the activity itself—the cooperation and killing. These considerations all have some relevance to hypotheses about human hunting, and I note them here, since they follow from the evidence, but I will discuss them later, when we consider human hunting specifically.

The behavior of chimpanzees in the Gombe situation is most revealing—and this is not the aspect most evident to either fieldworkers or theorists—as it shows that the research itself is an evolutionary, or at least an adaptive, problem for nonhumans; it shows us the conditions under which problem solving and intelligence develop. This is even more dramatically apparent in the laboratories and primate research centers of universities. There captive animals must cope, as individuals, with radically altered environments. They must adapt not only to altered physical conditions but to the conduct of an alien species, *Homo sapiens,* and, correlatively and more importantly, to the food problem. For these researchers work chiefly with food, which they withhold and convert into a scarce resource, make available in abundance, and proffer in catering to preferences. The animals' procurement of the food is contingent on their solution of the adaptive problems presented by, belonging to, and reflecting human beings. And this is true whether scientists seek to have friendly chimpanzees nearby and in a natural state for purposes of observation or whether scientists make food available when chimpanzees move symbolic counters about or sign. Thus their intelligence begins to develop an objective rationality. These animals *learn* to express it in human fashion because human beings *teach* them (using sign language, for example) or formulate the problematic conditions (such as banana boxes). As the environmental problematic takes

shape, their intellectual competence manifests itself in performance—slowly, by trial and error, in individuals.

Now if that competence is evident in chimpanzees and is
brought forth, however primitively, by an environmental problematic focused on food and diet, I propose that a similar competence may have been possessed by early hominids. In the wild, all
primates, and especially the anthropoids, show a broad dietary
competence and preferences within a wide range. These are the
grounds for elementary rational thought about environmental
objects, for both characteristics imply selection and choice, as
well as permitting a generalized adaptation. Nothing that we
know of prehistoric primates, including hominids, precludes us
from thinking that they did not possess comparable competences
and were not faced with similar problems of choice, selection, and
preference. If artificial and dramatic problems placed before
nonhuman primates by human primates show signs of producing a
development of intelligence in the former, then there is no reason
to suppose that a comparable but "natural" environmental problematic, extending over a vastly longer period of time, would not
make possible the development of such competence for rationality into a performance.

The point is that our examination of chimpanzee behavior in
the Gombe has served to isolate food and diet as the focus of our
attention. This is where we should look in the evidence of prehis-
tory for clues to the development of rational objective thought,
though we cannot say whether the food problematic for humans
involved maximization of preference, resolution of ambiguity, or
the solution of scarcity.

Early Hominids: A Digest

The subject of the discussion is prehistoric species immediately
ancestral to the genus *Homo* and the earliest members of this
genus. This corner of primate taxonomy is, to put it mildly, in a

state of flux as investigators in the field unearth new African finds that throw previous conclusions into disarray and push back the known time of emergence. Tobias (1973) suggests *Australopithecus africanus* as the immediate ancestor of *Homo habilis* as well as of *A. boisei* (East Africa) and *A. robustus* (South Africa). But the recent finds at Afar in Ethiopia and Laetolil have thrown out this classification, without replacing it with one on which there is common agreement. At first the Afar finds were thought to have belonged to the genus *Homo,* but Johanson has recently (1979) proposed a new species of *Australopithecus—A. afarensis,* from which *Homo habilis* evolved and from which *A. africanus* branched. Mary and Richard Leakey dispute this and apparently wish to identify the Afar and Laetolil finds at *Homo.* The revised time scale seems rather more definite than the taxonomy. *A. afarensis* has been traced back more than 3 million years; *A. africanus* about 2.5 million years; *H. habilis* 2.2 million years; *H. erectus* 1.7 million years; *A. robustus/boisei* 1.8 million years; *H. sapiens* about 250,000 years. It is clear that both *A. africanus* and *A. robustus/boisei* were contemporaries of *H. habilis* and that *A. robustus/boisei* was also contemporary with *H. erectus.* It is equally certain that whereas the genus *Homo* has continued until the present day, the genus *Australopithecus* became extinct about one million years ago. The extinction of one may well be related to the survival of the other.

Whatever the outcome of taxonomic debates, and whatever new finds may turn up in the future further to complicate and refine the dates and place of early man, it will suffice to state here that I am concerned with the environmental problematic facing the various species of *Australopithecus* as well as *Homo habilis* and *Homo erectus.* I assume that the hominid solutions to this problematic were propounded more or less experimentally by *H. habilis* and *H. erectus* and were and are the continuing concern of the genus *Homo,* which includes *H. sapiens.*

Obviously our evidence of the diet of *Australopithecus* or early hominids is not direct but only vague and circumstantial. C. Jolly has proposed that early man was basically a seed eater, and this hypothesis has generally been well received (C. Jolly 1970). Hominids were nevertheless competent to eat a wide variety of foods, and this competence encompassed meat eating (see Szalay 1975).

To develop his hypothesis Jolly used as his model the gelada baboons, animals known to have moved from a forest to a plains environment and therefore presenting a possible analogue for the original move of early hominids from forest to savannah. Geladas feed in a sitting position and shuffle on their bottoms to a new grass patch. Such a posture, it is argued, may well have established the precondition for uprightness, while the baboons' extraction of seeds by the precise use of forelimbs and digits might have led to selection for an early form of the human hand and the necessary hand-eye coordination. Some workers argue, however, that *Australopithecus robustus* became a specialized seed eater before extinction. It may be that this very specialization caused extinction, while the generalization of *H. habilis* promoted survival, because although he may have preferred seeds and grasses, he did not totally depend on them.

The archaeological evidence for australopithecine diet probably accentuates the meat component simply because bones fossilize more readily than plants. The remains found at Olduvai suggest a partially carnivorous diet; the meat might have been brought back to a home site and shared out (Isaac 1971:288–89). There is also evidence of a preference for waterside sites, which indicates that early hominids, unlike the apes, could not derive enough liquid from vegetables and fruits and that they consumed fish and shellfish as well. Edible plants would also grow in greater profusion near water. In some areas the remains of only a few animal species have been found, while in others there are many, mostly of small animals. This discovery is consistent with the general primate pattern of individual and group preferences and variation within a given range. The sites that show the most spectacular evidence of meat eating, and possibly hunting, are associated with *Homo erectus* and occur in the middle Pleistocene in the colder, more seasonal regions of what is now Europe— Ambrona and Torralba in Spain, Terra Amata in France. This fact leads Isaac to suggest that the presence of meat in the diet increased in significance as early hominids moved into the more temperate zones, where seasonality would have a greater effect on the availability of plant food (1971:293). But it also indicates a radical difference between *Homo* and *Australopithecus,* namely that the former is dietarily more generalized than the latter, or that whatever reasons impelled the migration of *Homo* from Africa to Europe led to a confrontation with environmental problems that in turn called for enhanced performance of the meat-eating competence. Isaac makes more or less the same point in summarizing his survey:

> . . . in Africa, which may well be representative of tropical and warm temperate zones in general, hunting has seldom, if ever, been in any exclusive sense the staff of hominid life. The archaeological record, such as it is, appears more readily compatible with models of evolution that stress *broadly based subsistence patterns* rather than those involving intensive and voracious predation. During the Middle and Late Pleistocene the geographic range of hominids was extended into cold temperate and sub-arctic conditions. This almost certainly led the hominids into new ecological conditions where protein foods had to be dietary staples. [Isaac 1971:294; emphasis added]

But there is a problem hidden in this statement: as far as we know, *Australopithecus* remained confined to Africa, hence to tropical, subtropical, and warm temperate zones. It was *Homo* that ventured out of Africa into the colder, more seasonal areas of what is now Europe and Asia. Why did this genus migrate from Africa, and how was it possible for it to do so? It is no good invoking tautological "explanations" such as a supposed migratory disposition or a "spirit of adventure." We must seek for an answer that is consonant with evolutionary theory and fits the facts we have. The hypothesis I suggest is that the genus *Homo*, for reasons I shall discuss below, came into competition with other species, including those of *Australopithecus*, for preferred staple foods (probably vegetable) and found the solution in evading the competition by migration. The pressure was not so much on wants as on tastes, as Hegel succinctly put it. This might have happened because of the very characteristics of the genus that put it at a disadvantage in direct competition. As a generalized creature it could take advantage of environments beyond the reach of more specialized and limited species, but it would have been the weaker opponent of species that were specialized herbivores or carnivores.

Some evidence in support of this hypothesis relates to dentition. Changes in hominid teeth began very early, "certainly by the terminal Pliocene and [seem] to have been well under way even 10 million years before" (Pilbeam 1972:155). They apparently evolved in response to a new type of masticatory behavior, powerful slicing anteriorly and crushing and grinding posteriorly. The result was the "reduction and incorporation of the canines into the incisor cutting battery and the development of thick enamelled flat-crowned cheek teeth" (ibid.). At the same time the shape of the dental arcade changed from a V-shaped to a parabolic curve evident in the fossils now designated by Johanson (1979) as *A. afarensis*. Such alterations have been interpreted as an adjustment to a diet of seed and grass, but they also have some bearing on the other functions of teeth in primates. In early hominids and humans, the canines and incisors no longer project forward from the rest of the face, and this, together with the fact that they are drastically reduced, suggests that certain earlier functions of these teeth have disappeared. In nonhuman primate apes, canines and incisors may be said to function more as *tools* and weapons than as masticatory devices. They cut, tear, pierce, and otherwise serve to extract material from the environment for use as food. The baring of the canines is one of the principal actions in the aggressive displays of primates, and these teeth are used, together with the limbs, as weapons in fighting. (In this sense they are almost at one with the limbs in their function.)

The reduction of the canines and their transformation into incisors in *Australopithecus,* and even more markedly in the genus *Homo,* indicates the loss of a major weapon not only against enemies but for "social control" in conspecific populations. Thus we can say that the *somatic* evolution of nonhuman primates in regard to their dentition and its alliance with the limbs has proved adequate to their environmental situation, while the total somatic evolution of *Australopithecus* and *Homo,* though enlarging dietary competence, has at the same time reduced the offensive armory at their disposal.

Similarly these dental changes reduce the effectiveness of the teeth as tools of extraction, while allowing for a more generalized masticatory ability. The development of early hominid dentition suggests a generalization of diet such that seeds, fruit, leaves, and flesh can be masticated with more or less the same efficiency. But like all generalized conditions, it puts a creature at a disadvantage vis-à-vis animals that specialize in eating seeds, leaves, flesh and so forth. Early hominids would not be as skillful as carnivores at killing, scavenging, and eating meat, or as clever as gramnivores at procuring and masticating seeds. As Szalay (1975) has pointed out, recalling an earlier work by Gregory, the incisivization of the canines in early hominids means an impoverishment of the somatic tool kit relative to that of other primates. Today, teeth still serve as crude tools among certain peoples (including the Australian Aborigines), sometimes acting as a vise or assisting in the stripping and planing of branches for spears.

But when the teeth in early hominids lost importance as tools and weapons, they also became divorced from the limbs, the forelimbs in particular. We might presume that the invention and use of tools—as extensions of the limbs that substitute for the teeth—was a corollary.

If, as I suspect, the range of dietary competence of *Au-*

stralopithecus and *Homo* in particular was broadening, then these creatures were likely to have been in competition with *all* other proximate primates for food, as well as with animals from different orders. It is unlikely, in other words, that early hominids lived sympatrically with each other or with other primates; more probably, whichever way they turned, they ran up against a competitor, one who more often than not was better, more efficient, or stronger within particular niches. At the same time, the increasing generalization of hominids and australopithecines allowed them to move from one niche to another, or to take advantage of the resources of a wide variety, though relative to any specialist they were at a distinct disadvantage. Hominids became the universal competitor, the perpetual nuisance, as they were motivated to move about.

Generalization of dietary competence, a response in keeping with hominid primate nature, enabled populations to expand their horizons and to survive against competition by seeking new pastures. Yet the consequence was a generalized somatic condition in which individuals were only fair to middling in matters of offense and defense. While hominids (and contemporary humans) can climb trees, they cannot do so as well as chimpanzees; while they may be stronger than monkeys, they are not as agile as gorillas; while they can run tolerably fast, a lion will reach an antelope long before a man will. Yet the genus *Homo* has survived, and the genus *Australopithecus* has become extinct. *A. africanus,* according to some taxonomists, was ancestral to *Homo* (see Tobias 1973) and might be said to have survived vicariously. How, then, was the survival, let alone development, of the genus possible in the face of broad capabilities in many environments and disadvantage in all?

Food and the Brain

The cornerstone of my argument is that *Homo* evolved as the most generalized of all primates in all respects—morphologically and socially, somatically and psychologically. But as with all generalities, there is an apparent exception. Tobias, quoting Stephan, writes: "Only man has encephalization which exceeds that of all other animals. He is the only primate with an outstanding brain size" (1975:384). This means that hominid species evolved the largest and most complex brain *relative* to body size. Other animals with larger bodies have larger brains, and even the large brain of *A. robustus* seems more a function of increased body size than of encephalization. This is only an apparent exception to the generality, however, for as I have explained earlier, a generalized morphology presupposes cerebral development.

A comparison of the estimates of body weight and brain size for *Australopithecus* and *Homo* gives a clear indication of hominid encephalization. Tobias's data (1975:365–368) may be summarized as follows.

A. africanus   had a suggested body weight of 40–50 pounds and a brain size of 428–80 cubic centimeters, with a mean of 441.2 cubic centimeters.

A. robustus   probably weighed about 150 pounds and had a brain size of 506–30 cubic centimeters, with a mean of 519 cubic centimeters.

H. habilis   is estimated to have weighed 50–60 pounds and had a cranial capacity of 593–684 cubic centimeters, mean of 460.5 cubic centimeters.

H. erectus   may have weighed around 120 pounds, and the combined brain size range for speciments from Africa and Asia is 727–1,225 cubic centimeters, with a mean of 931.5 cubic centimeters. The Asian specimens seem to have had a larger brain than the African, a difference that may be due to the greater complexity and changes encountered in the Asian environment.

Holloway (1975) has suggested that between Australopithecus and apes there are distinct differences in brain conformation, and that between *Australopithecus* and *Homo* there are differences that clearly indicate the development of a human brain. He writes:

> In sum, then, the whole brain of hominids . . . shows a number of reorganized features, mainly based on the expansion of the posterior and upper parietal cortex, expansion of inferior and anterior temporal cortex, more gyral and sulcal folding, and more modern disposition of the so-called "Broca's" area of the frontal lobe. [Holloway 1975:405]

This development of the brain, together with its enlargement, is not a process that would have occurred in a vacuum, for its own sake. These changes either arose as adaptive responses to problematic conditions of the environment or, if they did arise by saltation, were selected because they proved adaptive.

Since it is out of the question for the genus to develop somatic specialities, it can meet environmental problematics only by combining subjective reasoning with objective reasoning. In short, the general solution to hominid adaptation amounts to the development of intelligence, very much in the same manner and under the same problematic conditions as those experienced by Jane Goodall's chimpanzees.

Early hominids in Africa seem to have been primarily vegetarian and probably preferred grasses, grains, and seeds. But they could equally well have subsisted on leaves and fruits. Either way, any such preference for a staple would bring early hominids into competition with baboons (if they preferred seeds) or with the great apes (if they preferred fruits and leaves). Modern *Homo sapiens* is nowhere near as strong as a chimpanzee or as agile. It is of considerable interest that a large number of baboon bones have been found at australopithecine sites, the implication being that the latter killed the former.

Faced with proximal competitors who share the same preferences (and knowing that the genus *Homo* has come out ahead), we may reasonably argue that the difference lay in the competence of the more generalized genus to outwit the competitors. The brainpower is directed toward developing some means of superior force. It may be regarded as having two dimensions, the creation of weapons (possibly from tools) and the development of social organizations to overcome individual disadvantages.

But this is not the only avenue open for overriding competition. Early hominids may also rely more and more on eating meat, thereby relieving any direct competition with other primates, apes and baboons in particular. We recall that under similar competitive stress over staples, chimpanzees in the Gombe showed a distinct increase in hunting and meat eating, as well as aggressiveness.

As far as the procurement of meat is concerned, hominids are not especially well suited to killing, in comparison with other primates or, more particularly, with specialized carnivores. Chimpanzees kill with the hands and teeth, though they suck the meat rather than cut it. They are also fast enough and agile enough to catch their prey bare-handed. And while we may certainly grant that not all hominids are as unfit to hunt on two legs and with bare hands as modern Western man is, I would still assert that between the abilities of human and nonhuman primates there is a significant difference in weapons and in cunning. If hominids turned more and more to meat, then they would come more and more into direct contact and competition with predatory carnivores. At this point the hunter, man, if he was not actually hunted, may well have had to take precautions. Hominids are no more "naturally" hunters than they are fishers. They are simply competent to hunt, but to turn a dormant competence into a successful performance, early hominids would have had to develop and improve their ability to stalk, hide, and kill. They could surmount this problem to some extent by scavenging, but here again they would have to compete with numerous, far more efficient and specialized species, from vultures to hyenas.

None of these problems can be met with any great success by an individual hominid using his body alone—neither the limbs nor the teeth of hominids can serve as adequate tools and weapons in their own right for meat eating and hunting or, more emphatically, against specialized competition. The solutions have to be *extrasomatic*, namely, the development of strategic and tactical ways and means to secure food, to outwit opposition, and to gain permanent advantage. Since nonhuman primates placed in comparable situations quickly develop their competence for tactical thinking and organization when faced with problems of securing a preferred food, it is hardly too farfetched to suggest that early hominids performed in the same way. We can assert that early hominids had the capacity to develop such a response because we have the evidence of encephalization. Further, there is the circumstantial evidence of Oldowan tools and weapons found in association possibly with *Australopithecus* and indubitably with *Homo habilis* and *erectus*.

A third solution to this broad environmental problematic was migration. For some populations of *Homo habilis/erectus,* we can propose that the competition forced them to move farther and farther across the land mass of what is now Southern Asia (via the Arabian Peninsula). Or it may well be that for some early hominid populations, migration and exploration was a matter of choice. Thus far the archaeological evidence suggests that hominids emerged in Africa, thereby supporting Darwin's contention, and so presumably spread out from Africa to present-day Asia, where no fossil remains of *Australopithecus* have yet been found.

These three broad forms of solution are not, of course, mutually exclusive; any one, any two, or all three may have been adopted simultaneously.

The advent of meat and the attendant difficulties of its procurement have rightly been emphasized as important for human evolution, but it has not always been made clear just why. The present argument suggests two possibilities. First, the advent of meat indicates a breaking out of the vegetable or floral barrier, while a developing preference for it helps hominids adapt to geographically varied environments. It has been shown experimentally that animals naturally strive for a nutritionally balanced diet (see Katz 1953:122). Under the conditions of competition and uncertainty faced by early hominids, the new protein source would help balance reduced access to vegetable matter and supply increased energy for competition, migration, searching, and such physical activities as tool making and defense, all of which were becoming increasingly characteristic of hominid life.

Second, the procurement of meat is an activity that directs attention more closely to the nonhuman world. Since animals move, have wills of their own, are specialized in their senses, and can reason to some extent, they present a more complex problem and a more intricate subject of knowledge than the plant world does. If hunting is to be effective, then, it will require greater thought than would gathering; that is, the intelligence implied by gathering is extended by hunting.

Meat gradually becomes an increasingly regular feature of hominid diet as a result of competition. If Isaac is accurate in suggesting that meat eating assumes greater importance in the more temperate and colder zones, then it may be a consequence of migration.

In the Gombe reserve the overt aggression shown by chimpanzees, which led to predation and killing of baboons, was confined to males (Teleki 1973:35). It may be possible to argue that the procurement of meat is linked to aggression and has been a male speciality from the beginning. The physical disadvantages of hominids vis-à-vis other animals may be primarily sex linked; the usual primate pattern is for males to provide defense against enemies, to indulge in dominance displays toward other males, and to behave aggressively in provoking circumstances. Furthermore, among many primate species there is a marked sexual dimorphism, with males being larger and stronger than females. Such dimorphism seems also to have been characteristic of early hominids, or at least those found at Afar, Laetolil, and elsewhere. Indeed, classification is often confused by the possibility that different specimens belong to different sexes rather than to different species.

It is only fair to note views holding, for example, that contemporary hunters are by no means aggressive people and that hunting, for them, is far more a sympathetic or benign activity than an aggressive, malignant one (Fromm 1974). Still, intensified aggression, sexual dimorphism, and hunting/meat eating intercorrelate, and if the evolutionary problematic concerns *species*—or at least populations—survival then we may well ask how females and young acquire meat. If they do not make a habit of hunting themselves, then they must receive meat from males, and this may have been how the obligatory sharing of food originated. If we take contemporary hunters and gatherers as an example, we rarely find gathered foods necessarily shared, although they may be divided as a matter of course. Individuals, even young children, are capable of gathering for themselves, but formalized, almost ritual sharing appears only to apply to meat—and not to all meat caught by hunters. Nor do hunters always apportion their spoils, which suggests that the obligation does not come naturally but is imposed upon its members by a community.

If the sharing of meat is what we might call the elementary subject of ritual imposition focusing on objects external to the human (or hominid) being, and if, for the reasons suggested above, hunting and procurement of meat are a male concern, with females as the main beneficiaries (or those to whom the obligation is due), then we have some basis for recognizing the conditions that make possible the division of labor as a cultural phenomenon—as well as some reason to suggest why the formal attribution of power to males and their elevation in ritual is as it is. If we remember that the situation argued for here is coterminous with kinship and social organization, as described in earlier chapters, our thesis overall is a strong argument (though admittedly entirely speculative) for explaining the relative cultural and ritual status of male and female.

This situation must of course be regarded as having crystallized from attempts to grapple with the subtleties and variations of the environmental problematic faced by early hominids. The core of this problematic is the interrelation of food supply, competition for food, and food preferences. Food is the impetus for rational, objective thought. I will next consider this notion more analytically.

Men of Good Taste

All members of the Hominoidea, apes and man, show an eclectic taste in food but select, from a wide range of possibilities, only a few to provide the bulk of their diet. According to season and place, they show clear preferences for specific plants, leaves, fruits, and so forth, and they will generally eat only lush morsels of the best quality. Thus chimpanzees of the Gombe clearly prefer bananas to anything else, but before their advent in such profusion, the animals seem to have liked figs best of all. I have no doubt that it would be possible to determine orders of preference among populations of apes for specific food items.

The broader the dietary competence (and the greater the fertility of the environment), the greater the opportunity (if not the necessity) that presents itself for exercising free choice, and this exercise entails the recognition of criteria for selection. Such judgments must be based on the matching of *subjective tastes* (a sensory attribute) to perceived qualities of the *objects* existing in the environment. What has once been experienced as a good taste or texture must be transposed into a visual, tactile, or olfactory attribute if the tasting experience is to be repeated. The look of a banana (or its smell) or the appearance and aroma of a piece of chocolate become a piece of *information,* a sign or symbol of some aspect of banana or chocolate. This sign *separates* one food from another, whether it is a piece of bread or a fig or a leaf. On the basis of this received information, the animal can *formulate an intention,* impelled by his desire, from which he may *proceed to actions* that may or may not obtain the food—experimental or instrumental actions that are repeated on the basis of memories. This procedure will also entail the overlooking of other possibilities, such as the presence of other food. It was notable that, with the abundance of bananas, the chimpanzees in the beginning simply sat around Goodall's camp, waiting—though there was plenty of food in the forest. Realizing a discovered preference thus involves proceeding from a sensation through a perception to an intention, thence from an intention to an experiment that is aimed at producing a technique that will obtain the end first made explicit by the desire. Such is the procedure of reasoning that we have seen so skillfully demonstrated by chimpanzees in the Gombe reserve—but not only by animals in their natural habitat.

Sarah, the chimpanzee trained by the Premacks to communicate symbolically with plastic counters, was taught with rewards of food, and she showed quite definite dislikes and preferences, especially liking chocolate (Premack and Premack 1972). Washoe, taught by the Gardners to communicate in American Sign Language, seems to have had a preference for sweets and soda pop, though it is less clear how important food offerings were in her training (Gardner and Gardner 1969). At first the human world of scientists was astonished that an ape could learn a human symbolic language. Recently a more skeptical note has been sounded as critics question whether these apes generated grammatical sentences from their learned vocabulary or whether they simply responded with imitations of their teachers' questions and promptings, cued by involuntary visual signals given by the instructors, such as body stiffening and relaxing (Terrace 1979, Sebeok and Umiker-Sebeok 1979). Whether these animals learned human language or not, the evolutionary implication of these studies is that under circumstances of drastically altered environments, particularly those factors affecting diet and emotions, chimpanzees learn to use a novel and to them peculiar strategy to obtain food and comfort. Other animals in captivity obtain titbits, delicacies and emotional rewards by standing on their hind legs, counting, or jumping through flaming hoops; these chimpanzees perform signs and gestures that imitate human symbols. What these studies do show is that apes can respond strategically to environmental problems. Since these problems are humanly contrived, their solutions resemble human reasoning. "Nim's and Washoe's use of signs suggests a type of interaction between chimp and trainer that has little to do with human language. Nim's and Washoe's signing appears to have the sole function of requesting various rewards that can be obtained only by signing" (Terrace 1979:76).

It is precisely the broad dietary range of hominoids, and chimpanzees in particular, that calls for discrimination, separation and choice, suggesting that we should attribute to the chimpanzee a rudimentary *theoretical knowledge* of the world, a classification of the environment and its constituents, through which visual, tactile, auditory, and olfactory sensations may be transposed into taste sensations and through which intentions can be formulated such that selected actions may realize them. If chimpanzees did not possess this rudimentary knowledge, I doubt whether human beings would so readily recognize a kinship with chimpanzee behavior and disposition and be so ready to bestow the accolade of "intelligence" on them.

Chimpanzees (but not only chimpanzees) show remarkably humanlike abilities to learn human ways in laboratories, field stations, zoos, and circuses—but only because they are learning the "contours" of the environment, and this is a *human* environment. Human beings *instruct* chimpanzees (we tend to "teach" chimps, but to "train" horses or lions), and we teach them human ways, we transmit to them information from an environment. Because they are more or less generalized to a degree approaching the human condition, they are capable of learning from an environment (that is, they are not so completely specialized that they are irretrievably adapted to a particular environment).

The continuity of hominid with pongid thinking lies in a shared ability to rationalize about objects in the environment and about themselves in relation to those objects. Both orders have an ability to categorize and classify items of the environment as food and not food, as good food and not so good food, to define items of the environment in relation to taste preferences, and to establish some sort of canon for transformation of certain sensory information (visual, auditory, and tactile) into understandings of the other senses. In the light of all this, early hominids, like all hominoids, cannot be denied a rudimentary theorizing ability, a capability for rational or instrumental action by which methods arrived at by experiment may be repeated by individuals to secure the ends that satisfy their preferences. The earliest hominid, *Australopithecus,* would certainly be no less intelligent than any other primate. But this basal intelligence is little more than a form of operant conditioning in which the environment poses a problem that is solved by individuals or populations finding out how to obtain their rewards. The question suggested by the evolution of human intelligence is, How might this preadaptation have *developed* into a *permanent* and *self-generating* form of life that secures a degree of *independence* from environmental conditioning?

Chimpanzees have *learned* to communicate symbolically and to find out novel ways of obtaining their ends (unscrewing screws, for instance). But all this happens only when they are *taught,* either directly or indirectly, by human beings. Chimpanzees learn, but they do not teach. Figan, Fifi, and Evered could open boxes but appear to have discovered how to do so independently; other chimpanzees did not figure it out, and Figan and the others did not (could not) show them. It remains to be seen whether Sarah, Washoe, and the countless other victims of human egotism are able to teach their progeny sign language or some other symbolism. My guess is that given a sufficiently long perpetuation of their changed environmental conditions (an unlikely possibility), they might very well come to do so. But even if they eventually prove themselves thus capable, that is, even if chimpanzees *evolve* toward being more humanlike, this will not tell us much about how human teaching and symbolizing became possible. For while chimpanzee adaptation is achieved through learning, imitating, and trial and error experimentation, human adaptation is achieved through teaching as well. Chimpanzees and other apes adapt, make choices, theorize, and classify via *direct* "instruction" from the environment and their experience of it (see Gibson 1966), but human beings learn not only directly from the environment but also *indirectly (about* the environment) from teachers. A human child, for example, learns not only by experiencing the environment directly through reading environmental information and transposing it into its own sensations but also without such experience or through a mediating experience. Taught information is isolated, *separate* from its context. Teaching (as distinct from learning) means a *form* of technique and knowledge that is independent of the content of the form. For example, to teach someone *not* to eat a certain plant, the teacher cannot eat the plant and make some sign of negation. What is to be taught must be modeled or represented, as it were. Thus a preliminary answer to our question is that teaching is the activity that transforms or develops an elementary intelligence into a permanent, self-generating, and evolving form of life. Now the question is, How is this activity possible?

This form of the question returns us to the matter of subjective reason, for from a species point of view, teaching is a subjective activity—members of the species instruct each other about something outside their confines. The development of objective rationality (but not its origins) rests upon and is presupposed by the development of subjective reasoning, though the one does not necessarily happen before the other, and indeed, logically, their development is a matter of interdependence and feedback. To account for the possibility of teaching, then, I need to return to the earlier parts of my argument.

The crucial feature of the earlier, social argument that is most relevant here is that an autonomous social structure is possible when determined relations (pair and primary bonds) come into relation with each other to produce a third and new relation (father). This new relation can be perpetuated, however, only if the original elements that generate it are retained and can be reproduced. Thus the possibilities of larger structures made possible by the transfer of relations through the father must be balanced by mechanisms that will allow the autonomy of elementary, determinate relations, and not simply as a structure but also in terms of their function.

As we have seen, relations that are joined for certain purposes must for other purposes be separated. In our paradigmatic model, if the pair relation is to be realized to the full, then the primary bond must be held in abeyance. If the woman is paying attention to the man, then her attention to the child must be allowed to diminish, even if it is only momentarily.

This situation is manageable only if some form of representation is made, assuring the existence of the relation held in abeyance by conveying the sense that at some future time, it will be given priority. I have called this assurance the promise. But sine the promise is of a *future* performance, it is impossible to convey it directly or through an immediate analogue. The promise must be conveyed symbolically, and its immediacy to the structure and function of human relationships leads me to suggest that it is the origin of human subjective reasoning in its symbolic form. If the promise is to work, it must represent convincingly something (a future action) that is not there: it must be given and taken as an earnest, a symbol of intention and intension, that is, of the existence of the *idea* of the relation between desire, purpose, means, and ends and of the extension of the internal appreciation of the self to encompass another person. If there are creatures who make promises to each other, then it is possible for them to teach each other, for teaching is an activity that rests on forms without reality beyond themselves and prepares the learner without requiring him to experience the object taught. In the Caroline Islands, for example, an extremely complex mode of navigation is practiced that allows islanders to travel hundreds of miles across an empty ocean without the benefit of navigation charts, compasses, or indeed, any instrument. Such navigation cannot be taught by allowing the novice to learn from experience, but those who know teach through designs made in the sand and by telling highly structured stories. The method of teaching is not the method of navigation, yet it teaches navigation (see Gunn 1979).

There is, then, an art of teaching that is independent of what is to be taught. The designs and stories are about navigation but are not *of* navigation. Rather, they are of teaching—products devised to get ideas about navigation across. Thus we have another generalized accomplishment of the species, a technique that makes it possible to produce numerous different forms, whether navigation, agriculture, etiquette, or language.

Teaching may be likened to the machine whose purpose and function is the production of tools that will be used to produce specific artifacts. The discovery or invention of teaching makes possible the reproduction of the specific techniques that in their turn produce the numerous items of a culture. Thus the art of teaching is used not to make tools, but to transmit the technique for making tools or for adopting a kinship system or for participating in ritual. It is the development of the art of teaching as an art (or a science or a skill) of culture as a whole that seems to me to be a significantly human accomplishment. And from this point of view I would amend the cliché that culture is learned behavior; rather, culture is taught behavior. For the evidence is overwhelming that all living species learn or are capable of doing so within limits and that some species, the nonhuman primates, for example, can learn very effectively, but that the development of teaching reaches its highest extent in the human species.

This development is it itself made possible only through the emergence and development of human subjectivity, the fact that individual human beings can and do take account of each other in their activities and can and do use each other to mediate between an individual and his environment. Teaching rests on the capabilities of human beings to establish and determine for themselves forms of relatedness and to reproduce those forms. This reproduction entails a vision of the future such as that embodied in the idea of the promise, and the promise is that which makes possible the ability to represent forms in some other medium (that is, symbolism) or to represent forms that do not otherwise exist (because, for example, they will become actual in the future).

CONCLUSION

The Darwinian theory of evolution is part of the evidence for the apparent uniqueness of humans. For what other species has demonstrated its interest in the development of life and, on the basis of its discoveries, has attempted to transform life? Yet this manifestation of human originality contradicts the central tenet of the theory of evolution by natural selection, "Natura non facit saltum." If nature does not make leaps, then there can be no clean break between the human and any other species.

When it was first formulated, the theory of evolution was regarded as a refutation of the prevailing ideas about creation by design. Darwin's case for the gradual accumulation of "slight, successive, favourable variations [that] can produce no great or sudden modifications [but] can act only by short and slow steps" proved so overwhelming and so consistent with the facts that the design argument had to bow out—except, that is, with respect to the evolution of the human species. For man, the anatomical evidence of slow and gradual modification seemed in accord with evolution, but the gulf between what humans and nonhumans could say and do and think seemed too great for evolutionary continuity to carry the day. It is not difficult for anatomists and paleoarchaeologists to accept the gradual evolution of hominids. Existing gaps will be filled eventually, if and when the missing fossils are located. But for the evolutionist who considers human thinking powers—the "mind"—patience holds no such hope; the

148

divide between the human and the nonhuman seems absolute and real. Still, as long as we argue that there is "something special" about the human species—be it language, rationality, powers of mind, symboling, ultimate concern, planning, choice, or reflection—then we contradict the theory of evolution itself.

Darwin, in *The Descent of Man*, noted the problem and asserted, "Everyone who admits the principle of evolution must see that the mental powers of the higher animals, which are the same in kind with those of man, though so different in degree, are capable of advancement" (1899:609). There is no reason to suppose that such development cannot be accounted for by natural selection. Why, then, have so many thinkers since Darwin—individuals whose scientific eminence is undoubted, such as Lorenz, Dobzhansky, White, Waddington, and Monod—argued for the special significance of special traits special to man that have had a special origin?

I suggest that at least part of the reason is the conflation of the time factor in evolution: present human accomplishments are regarded as being of the same order of magnitude and distinctiveness as the feats of the species at the time of its emergence. Because today human thinking appears so far removed in scope and complexity from that of other primates, it is tacitly assumed that there must always have been such a gap. It is taken for granted that human distinctiveness was there from the beginning and therefore that it must have been specially created—the result of a miracle (Monod on language) or a creative flash, a spark of cosmic ignition: "Each step forward has consisted of a *fulguratio,* a historically unique event in phylogeny which always has a chance quality about it—the quality, one might say, of something invented" (Lorenz 1977:35).

What I have sought to show in this essay has, I hope, remained faithful to Darwin's theory, namely that the species nature of *Homo sapiens* has itself *evolved* through the natural selection of variations occasioned by the changing conditions and circumstances of the total environment, which continually confronts the human, and indeed all creatures, as a problematic. Although the scope of human accomplishment and capability was not as broad in the Paleolithic as it is today, it is nevertheless the result of the accumulation over time of "slight, successive, favourable variations." I have tried to show this by shifting the focus from the development of man to the conditions making it possible, so that evolution is considered not from the viewpoint of the species but from some point on the edge of the totality that this evolution comprises. In fact, as will be readily seen, I have spoken about the emergence not so much of human "culture" as of the conditions of a totality that invite this culture as a circumstance. My attention has been on a moment, the Plio-Pleistocene era, and, simultaneously, on a context, the primate order. Man's development is thus seen not as a subject of Evolution considered as a whole but as a function, as a process that does not itself distinguish, as man does, between one species (a subject), seen as evolving in relation to a separate environment (an object). In a sense the present essay tries to provide an epistemological basis for a theory of the emergence of human culture and human society.

It can be argued that whereas Evolution displays only a past career, human development may be considered teleological, its purpose being defined within Evolution, which is no more than a description of the events of natural selection, the survival of variations that interrelate differently over time. But since it has become evident after the fact that human nature and its skills have evolved to the extent that a huge gap has opened between the human species and the rest, and since this gap has made such a *difference* to the variants that comprise the world in its environments, might it not be argued that Evolution has in some way made the difference? At least to the extent that the human species can *now* operate with purpose in mind? That through the human species *Evolutionary purpose has itself evolved?*

There is little doubt that human individuals and populations can and do act according to their ideas of the goals they have in mind, that the difference between the most skillful bee and the most incompetent of architects is that the architect works from a plan. This planning or sense of purpose is itself a species characteristic that has emerged as a part of the way in which we think about the world and act within it, hence of the way in which we think about and act within Evolution. So as the sense of behaving with purpose has evolved in the human species, it has become a means to structuring events around us and within us—and this includes Evolution. Evolution, from a human point of view, can be *made to show* an aim, and our ability to make it do so belongs to both the human and the larger process. As far as we are concerned, a conscious sense of purpose is a variant selected for because of its adaptive advantages for us as a species. It pays us to consider human evolution as teleological, and since we are part of the greater whole, to consider Evolution as being in *some sense* teleological.

When we study the history of the current state of affairs, it is virtually impossible for us not to assume a sense of direction, since we are inevitably working back from an end point (the present) to some sort of beginning. It is a simple step to take the Aristotelian view, that the conclusion of any process comes before its beginning as a matter of fact, for otherwise things would not have turned out as they did. The very fact that we must start somewhere when we try to describe events in words involves the notion of purpose.

Thus the present essay argues that the evolutionary characteristic of the primate order is a tendency toward increasing generalization of morphology. Given that the human species is the most generalized of primates, it is hard to escape the conclusion that we represent an end point, as of now. We cannot say this is an absolute end point, except insofar as the conditions of generalization are those that make purposeful acting and thinking both possible and necessary.

It seems quite unexceptional to claim that primate and hominid generalization developed simply as the result of natural selection. This is an order whose members are selected for their flexibility, as opposed to other orders, whose members' variations are selected for their specialized adaptability.

But generalization, in principle allowing the most flexible individuals to take advantage of any (or most) contingent environmental situations, contradicts the very principle of natural selection, which concerns *variations* that can only be of a specific nature. Since the environment comprises multitudes of species selected on the basis of such differences, it is itself specialized. Yet the primate order in general, and the human species in particular, show no specific variants that can be selected to fit the organism to a setting.

The question is whether this contradiction is "real" or simply semantic. It seems real to me; it can be tested against the evidence of experience. Comparative anatomy displays the generalized nature of the human frame relative to the skeletons of other species. The diversity of human cultures, whether looked at through time or at any given moment throughout the world, is equally solid evidence of human generalization. And unquestionably the different parts of the earth, containing its myriad species, present a huge mosaic of different and particular settings.

It is from this real contradiction and the need for its resolution at its various levels, by every human population and every individual, that the sense of purpose has evolved. For only when an organism of general abilities confronts a contextual situation as a problem is it presented with any idea of a purpose. The organism cannot guarantee that what it does to produce and reproduce will succeed, since it must guess at the best way to achieve these ends. Such reasoning entails the separation of the goal from the means of attaining it and brings a corresponding doubt about the goodness of fit. It must hardly occur to a giraffe during the normal course of its life that there is any difference between reaching up to the higher branches and getting leaves to eat. But for humans there is always a discrepancy between setting off on a hunt and coming back with meat; between planting seed and harvesting grain; between looking for berries and roots and finding them.

The human infant comes into the world prematurely, as part of the condition of being generalized, unformed into anything specific. But from birth it must face the problems of the species, adapting its dispositions to the shapes, forms and dispositions of the environment, which include the bounded contours of the body itself. It is therefore a condition of human nature that the individual confronts the contradiction of existence at all levels, so that, as Lacan argued in 1958, the first problem is to make sense and to solve the contradiction of the body's specificity in the light of the generality of perceptions. For Lacan, however, this reconciliation is the primary problem in an evolutionary sense. But I am asserting that the resolution of the conflict between body and feeling, which produces the idea of the "self," is but a transformation, an entailment of the phylogenetic contradiction of the human species in its world situation.

It is in this sense that the ontogeny of the individual recapitulates the phylogeny of the species, since each individual must cope with the species problem of adaptation. Yet in a way this problem is not unique to humans, for as Lorenz's experiments have shown, and as any farmer knows, an animal's "idea" of itself is gained through imprinting, as if it saw itself as a reflection in and of those organisms to which it was originally exposed. By virtue of our generalization, the human problem of adaptation and identification is both more intense and more wide ranging, the complexities of environmental specifics being far greater. Consequently our situation in evolution becomes more focused on a contradictory state of existence. Its exact form must vary according to the historical circumstances of the population and of individuals, but all versions show similarities as expression of the general condition of opposition in which the species as a whole has been placed. Conceptualizing the process in this way makes its reality vivid and immediate, thereby providing a motive for action, a prescription for conduct. Thus thinking and its expression become facts of life and part of the human response in evolution.

Each individual, by virtue of his humanity, thus faces throughout his lifetime a problem of establishing and maintaining his identity. In Maori society, whenever a person wishes to make his presence known to others, to express his opinion on matters of concern, or to establish his right to partake of the resources of land and people that will permit his survival, he must establish who he is. If a man intends to speak on the *marae* (place of assembly), his right to do so may be challenged:

"Ko wai to Maunga?"   (Who is your mountain?)
"Ko wai to Awa?"   (Who is your river?)
"Ko wai to Iwi?"   (Who is your tribe?)

Unless the person can prove his identity by reciting his genealogy (*whakapapa*) and showing that he is *turangawaewae* (a man who has a right to belong somewhere) he is dismissed as a nobody— *enoho*. And before he can speak at a particular marae, his genealogy must show that he has this right. It is significant, however, that the identity of a Maori person cannot be assumed but must be established and constantly reaffirmed, and that the form of this identity is a man's adaptation to a natural and a human environment. He must show that he is a part; before he can fit he has to have been fitted.

It remains now for me to suggest where my ideas fit with existing theories about human evolution. The most straightforward of these theories deal with taxonomy and propose the sequence and relationship of the various fossil remains unearthed. These are the most literal hypotheses of human evolution, providing us with a genealogy, beginning with *Dryopithecus* and *Ramapithecus,* working their way through *Australopithecus,* to the species of *Homo.* Such reconstructions rely upon morphological and particularly anatomical features. They give us a picture of the earliest humans and provide minimal and often ambiguous clues to character.

The taxonomic theory of human evolution gives way to cultural theories. The moment artifacts are found in association with fossil hominids or in geological circumstances coeval with them, the theory of human evolution concentrates upon the objects as an index of the way of life of their makers. From such circumstantial evidence inferences can be made about the nature of early man. By the time of the Paleolithic, human evolution is cultural evolution.

Between these two forms of theory there lies a gray area described more by fantasy than theory, concerning what certain of these early ancestors "suddenly must have done." I would like to think that the present essay makes the "suddenly must have" theories obsolete. I have sought to offer a reasonable account of how and why specimens of a certain anatomical conformation, when confronted by various conditions, found it possible and maybe necessary to secure their survival in activities that produced tools; to assure their reproduction in relationships of kinship; and to protect the continuity and adaptive development of these activities through the exercise of thought. This is the transition from taxonomy to culture.

Second, the present essay links cultural theories of evolution, in which the conditions and constraints associated with the materials of a productive way of life are said to impose their own logic, and genetic theories of evolution, in which the conditions and constraints of genetic programs, particularly those constituted within and by the brain, impose their logic. Cultural theories see the unfolding of an historical, material logic that works itself out through human beings rather than in them. Genetic theories see human culture as the material expression of what is held in a state of immanence within the individual and therefore outside history. The very universality of the genetic aspect, not only for humans but for all animals, makes it inadequate to account for the particular evolution of a given species, especially one so distinctive as our own. But the logic of culture floats in a no-man's-land until we know just where it connects with biology.

I have sought to show that the genetic programs contained within the morphology of the human species are generalized, activated and shaped to specificity by the conditions, constraints, materials, and organization of the environment, including the very generality of those programs. Their logic must meet and adapt to the logic of shifting states of the world. In one form, the programs take shape as rational, objective thinking or "science." In another form they assume the guise of products—artifacts wrought out of the world's materials within the world's constraints. Our thoughts and activities, be they engineering or pure hypothesis, are a merging and a compromise between the logic of the arrangement of the world and the logic and arrangement of our genes. This procedure is arranged and instituted by natural selection—ideas and artifacts that work survive, as do the individuals who create them.

Finally, the present essay bridges a gap between the human and the nonhuman. The general theory of evolution insists upon the continuity of all species, and nothing in modern knowledge gainsays this (in fact everything supports it). DNA is DNA, whether it is in a human or in a jellyfish. At the same time our life experience and our deepest thoughts convince us of a startling and definitive difference, a hiatus between the two. Between the extremes must lie the ground of transition, which consists not in absolute fact but in how we have come to conceive of this continuity and this gap simultaneously. I have tried to show how the totality of circumstances involving the human and the nonhuman primate is such as to bring forth gradually differences of thought and action to such an extent that they make our privileged awareness of human uniqueness concomitant with the stark perception of animal continuity and unity.

Missing links, when found, tie together things that otherwise remain separate but not without their own integrity and validity. To the partisans of a sociobiological theory of human evolution, a cultural materialist theory of human evolution, an empiricist theory, a dialectical theory, a behaviorist theory, or a rational critical theory, I do not cry a plague on all your houses. I offer the grounds for an alliance.

APPENDIX: The Metaphor of Kinship

In chapter 3 I took for granted that kinship is universal to the human species and that it is always a means (though not necessarily the first or only one) whereby the relative identities of individuals are established. Still, it is impossible to verify this universality, in the sense that we shall never know as a matter of fact whether all human populations in the past acknowledged relatedness through kinship. The grounds for the assertion are partly intuitive and partly that thus far a population without kinship has not been reported.

There are some anthropologists who maintain that certain societies do not have a kinship terminology or a kinship system that can be isolated as such. Instead, it is argued, they are structured according to a system of prescriptive marriage alliance between groups that comprise relatively defined categories of persons, of which the basic divisions are wife givers and wife receivers. From this primary practice, it is suggested, the terms of address and reference are derived, serving not to establish individual identity but rather to indicate the association of persons in a category.

This argument, associated with Lévi-Strauss (1967) and Needham in particular (Needham 1962, 1964), reflects a deeply rooted theoretical viewpoint that was established philosophically by Hegel and sociologically by Marx and Durkheim. It asserts that the totality (in this case, the human social group) is logically prior to the individual and that the form of the totality provides the specifications for the definition of phenomena in general. This view opposes the individualist argument, associated with the British empiricists such as Hobbes, Locke, and Hume, which argues for the primacy of the individual and individual relationships, for their extension to form a group and, in general, for the epiphenomenal status of group structure—it is suggested, in other words, that the totality emerges as a consequence of aggregation and to a large extent without intention.

As a consequence of this opinion that in certain key instances the definition and arrangement of groups (for categories) is logically prior to the identification of the individual, Needham has recently argued that although anthropologists make the study of kinship a central concern, it cannot be shown that the object of their study exists. For in many instances, notably among those societies that practice prescriptive marriage, kinship terminology and behavior emerge epiphenomenally from the structure of groups operating within a system of alliance. "Kinship," Needham writes, "does not denote a determinable class of phenomena or a distinct type of theory. We are tempted to think it must have this specificity, because it is a substantive and because it is an instrument of communication. But it has an immense variety of uses, in that all sorts of institutions and practices and ideas can be referred to by it. . . . The term 'kinship' is what Wittgenstein calls an 'odd-job' word" (Needham 1971:5).

It is certainly true that the term "kinship" is an odd-job word, but does this mean that it does not possess a primary sense that describes something we are able to isolate? If we accept Needham's claim, at least in its literal sense, then one may well ask what chapter 3 is about. If there is no such thing as kinship, how can we discuss the conditions making it possible?

The basis of Lévi-Strauss's theory, which posits the origins of kinship in forms of marriage alliance operating as exchanges, is the incest taboo. It refers particularly to brother and sister, who are compelled to renounce their sexual rights to each other and (in order to ensure that they have a sexual partner) to exchange each other for one who is not a kinsman. Yet as I have already noted, before our hypothetical, primordial ancestors formulate— let alone observe—an incest prohibition, they must define themselves as "kin." If the groups of wife givers and wife takers engaged in exchange are themselves the imperative outcome of the incest taboo, and if this taboo can only come into being following a definition of kin, then it is obvious that the groups rest on kinship, and not kinship on the group.

In his pioneer work on social classification, Durkheim argued that the classification of natural phenomena was based on and extended the social division of men into categories and groups. To a great extent Needham has fought hard to sustain his view, though perhaps not as literally or naively as Durkheim. He has argued in numerous articles, and recently in a short monograph (Needham 1979), that there is at least a distinct isomorphism between classification of the world and the classificatory arrangement of groups that constitute the totality of a society.

In the present context, the question of the status of kinship, I wish to make another point. The Durkheim/Needham thesis proposes that social groups provide the structure for the symbolic classification of natural phenomena, and inasmuch as kinship is one form of such classification, it is argued that social groups are the model for the symbolic designation of individuals as categories of kin. This notion, however, implies that somehow or other human groups either form naturally or are delineated on the basis of wife giving and receiving, a definition that rests on the recognition of the incest taboo. We have seen that the latter possibility implies that kinship already exists; for the former alternative we are not given any indication of what criteria apply to the human social group that offers the basic, original precedent for further symbolic classification. There is certainly no evidence that mate exchange, with all the formality that it implies, exists as a naturally occurring criterion for group definition in either nonhuman primate or human populations.

A human group has definition resulting from the employment of criteria of inclusion and exclusion. It does not matter whether the category encompasses all males with protruding teeth or all persons on the mother's side; a *form* is derived *from* criteria, which precede the group. A classification or category cannot exist either as a physical entity or as a conceptual figment unless there is a prior identification of significant features of individual members. It may be said that certain animal species live in groups with particular forms which we call herds, troops, gaggles, flocks, and so forth—that may be hierarchical, single sex, multiple age, or whatever. We can say that animals make "natural" groups because individuals assemble themselves in this form *instinctively*. But we cannot say the same of humans. They live in aggregates that have particular shapes and structures, which in turn depend on significant features utilized as criteria for definition. There may well be a statistical predominance of certain types of group in the human species—all-male hierarchies or kinship clusters may well outnumber age sets or celibates—but all have form, are particular, and therefore owe their existence to consciously identified characteristics that reside in individuals or are thought to do so.

There is, of course, nothing to stop people from practicing economy and applying the same criteria to different domains; by extension and metaphor, humans may bring to natural phenomena criteria that define human groups. This, it seems to me, is precisely what Lévi-Strauss and Needham in their studies of symbolic classification have shown to be the case. What they fail to demonstrate is the primacy of the group over the individual.

The present essay has important bearing on this argument. If it could be shown that there is a logically and historically original *form* of human grouping, then it could be argued that the group antedates the individual, who must take his identity from it. I have contended, however, that this cannot be shown. To the contrary, it is a distinguishing feature of the human species that although individuals are naturally impelled toward living in company and it is necessary for them to do so, there is no universal instinctive *form* that this company will take. The options are open and depend on the conditions of the environment as they might exist not only "objectively" but also in the variable perceptions of individuals and populations. Thus men's social groups, as indeed all human classifications, must be constructed by the intention and will of individuals and according to criteria of their own choosing. To accomplish this, all manner of means are open, including the use of the natural grouping of animals as a model for people.

In 1971 Scheffler and Lounsbury published their detailed examination of kinship among the Siriono, who were a test case for the Lévi-Strauss/Needham thesis of prescriptive alliance systems rather than kinship systems (or at least kinship terminologies) and also provide an example of Needham's generalization that there is no such thing as kinship (1971). I doubt that even the most impartial reader of Scheffler and Lounsbury would agree with Needham that the Siriono have no kinship terminology, that they practice a system of prescriptive alliance between corporate groups engaged in exchanges between wife givers and wife receivers. As Scheffler and Lounsbury show, the Siriono do not have social groups of the corporate descent type; rather, they live in endogamous bands that are organized into a small number of matri*local* (not matri*lineal*) extended families that are not in any sense corporate (Scheffler and Lounsbury 1971:27). The sole basis for Needham's argument turns out to be, not the ethnographic evidence of corporate, lineal descent groups, engaged in a prescriptive exchange of women, but the *kinship terminology itself,* from the characteristics of which Needham *infers* the very social structure that he then asserts to be prior and to be the only real framework for classification in Siriono society. In this specific case, where it is claimed that groups are prior, that the alliance of groups generates the relative identity of individuals, and that therefore individuals are not identified according to a kinship system, such groups and such categories of person prove not to exist, but a kinship system does.

This is not the place to burrow further into this argument, and it is pointless for me to rehearse what Scheffler and Lounsbury have shown so eloquently, scrupulously, and convincingly. But I do wish to focus on their suggestion that kinship terms, thought of as monosemic by Needham and others, must in fact be recognized to be polysemic. This polysemy may be achieved in many ways—by extension of meaning and signification, by intension (or narrowing), and by metaphor. Indeed I maintain that the entire concept of "kinship"—which includes terms, ways of conduct, definition of rights, status, and duty—is metaphorical. Its metaphorical power in particular explains kinship's role in systems of classification, including the assignment of human beings to categories. Metaphor is one, if not *the,* primary mode of human thought, for in Shelley's words, "(Metaphor) marks the before unapprehended *relations* of things and perpetuates their apprehension" (quoted in Arendt 1978:102; emphasis added). Kinship is that mode of thought and expression which is concerned exclusively with relations. It is the metaphor for metaphor.

Needham quotes Wittgenstein in support and illumination of his contention that there is no such thing as kinship. But as a word it belongs to language, and in linguistic terms it is polysemic in English, that is, it may denote more than one object or have several meanings, which is all that Needham has in mind when he calls it an odd-job word. Because no one-to-one correspondence exists between the word "kinship" and any single thing or function outside language, Needham concludes that there is no point in looking for a peculiar something, a singular essential thing, that "is" kinship or in essaying a definition. But if the term is polysemic, it is perfectly in order for us to consider the possibility that there is a primary referent and that other meanings are extensions, intensions, or metaphors.

Reverting to Wittgenstein's discussion, we could also ask how we come by these useful odd-job words and what their nature is. Wittgenstein, in the *Philosophical Investigations* (1962), offers an illuminating though somewhat tortuous answer that at the same time solves Needham's problem and unwittingly illustrates the primacy of kinship as the metaphor, by which we seek to make known, or at least comprehensible, that which is unknown.

Wittgenstein asks us to consider the proceedings we call "games." He tells us not to say, "There must be something in common, or they would not be called games, but rather to *look* and *see* whether there is anything common to all." He concludes that this examination will reveal only a "complicated network of similarities overlapping and criss-crossing: sometimes overall similarities, sometimes similarities of detail" (1963:66). Wittgenstein then suggests that the best way to characterize, describe, or apprehend these resemblances is to say that they are "family resemblances."

Now what is accomplished by this attempt to clarify for us the puzzle of the variable similarities between things that are also dissimilar, such as games? Wittgenstein has used a metaphor and, what is more, the metaphor of *kinship*. He suggests that odd-job words like "game," "language," "meaning," and so forth refer to the relations between overtly dissimilar things among which there appear to be vague and elusive similarities, in the same way that resemblances are apparent between family members who also, in many ways, are quite different from one another. This well-known concept of Wittgenstein's—family resemblances—is itself a metaphor, and it is quite clear that games are not kin; the primary signification of kinship, the relatedness between persons founded on genealogy, can be extended only metaphorically in this argument. As Wittgenstein explains the odd-job nature of the word "language": "Instead of producing something common to all that we call language, I am saying that these phenomena have no one thing in common which makes us use the same word for all, but that they are *related* to one another in many different ways. And it is because of this relationship, or these relationships, that we call them all 'language'" (1963:65).

Kinship and family are equally odd-job words, but as I have shown above, they have a tremendous explanatory force, so much so that they can be used to make vivid and comprehensible comparable words such as "games" and "language" and "meaning." It is quite clear that in trying to convey the intricacies and subtleties of such a classification as "games" by comparison with "family resemblances," Wittgenstein assumes (quite reasonably, I think) that the latter is more familiar, more immediate, and more fundamental in our experience and in our thinking. Furthermore, the use of the analogy to explain what is otherwise inexplicable also supposes, quite correctly, that we have a rough analytical understanding of the basis of family or kinship resemblances—that they ultimately occur because of relatedness through birth, or more precisely, through the fact that family members are those who share and pass on genes. We could say that in other societies that are ignorant of genetics, people understand family resemblances because they are aware that in procreation people pass on blood, bone, flesh, or spirits by reason of the union of male and female in intercourse.

I have used Wittgenstein as my illustration only because the position of theorists (Needham in particular) who argue for monosemy and for the primacy of the group seeks to ground itself in his thinking. And his thinking itself, albeit unintentionally, provides the means to refute this argument and to support the alternative. But it should be noted that any number of illustrations can be taken as the starting point, including the fact that numerous societies classify natural phenomena in terms of kinship because this notion provides the major metaphor by which a relatedness between different phenomena can be grasped and communicated. It institutes something "real"—connectedness— where the information that comes to our senses does not do so. Thus "kinship" may not refer to a tangible something, as do other substantives, such as "table," "mountain," or "cow," but it denotes a mode of thinking about relatedness.

"Kinship" is a metaphor for metaphor. Still, as a metaphor, can it not be said to have a primary referent, from which other meanings extend? When we use the term, do we not have in mind something primary to which it refers, even though on the occasions of its metaphorical use we do not have this primary connotation literally in mind? If we describe the earth as "mother," we allude to a particular resemblance between earth (its bringing forth of plant life) and a female giving birth to an infant. But in mentioning mother earth we do not, I think, have human parturition in the forefront of our minds.

If "kinship" itself is a metaphor, then as a term it is comparative—it describes relations by analogy to something else, which in turn may not be called a metaphor. There has been much confusion about this fact in anthropological thinking in particular, for a lot of ink has been spilled in attempts to account for kinship as a biological, hence irreducible, phenomenon. And further, more confusion has arisen as the term has come to be used by ethologists and sociobiologists to refer to organisms that share genes. If "kinship" describes an irreducible biological fact, then it is not a metaphor; but if this is the case, then we cannot explain why kinship and related terms refer also to facts whose biological basis is neither evident nor posited by users of the concepts. The facts in question include intercourse, conception, parturition, insemination, suckling, and so forth, and their very existence in most if not all languages would obviate the need for kinship terms to serve such denotative purposes.

The word "kinship" does refer to relatedness between persons between whom connection is not physically self-evident but is posited on analogy with a self-evident and irreducible link. The relatedness, then, is both like and unlike a natural, sensual connection. The natural, irreducible connection between two separate human beings is the tie between a female mammal and her infant, evident in birth, in the placenta and the umbilical cord, and then in the close emotional and physical contact of breast-feeding. This tie is obvious and may serve as a basis for positing other connections. Such other relations are thus extensions, models, posits, or metaphors, for they are all reducible, as propositions of similarity, to something irreducible.

Kinship, then, refers to those links between persons that are postulated as *analogies* with the female/infant bond, and *not* to the female/infant tie—at least in the first place. The primary posited relationship is that of the adult male to the offspring of the adult female, and this is the first kinship link, as distinct from the biological model for the connection. It is in this, the institution of the "father," that we first come across variations in definition. I do not know of any culture in which the identification of a child's mother relies on anything but the biological evidence of birth; rather it seems that in all human cultures the female who bears an infant is acknowledged as having an irrefutable and irreducible tie (even when the child goes out for adoption, the "real" mother is still distinguished from the "adoptive" mother). But the positing of the father relation varies in human culture: a man may be "father" because the infant resembles him, has the same blood group, and is shaped and fed in the mother's womb by intercourse, or because this is the man living with the mother at the time of birth—and so forth. This variation is indicative of the metaphorical nature of the father relationship—that the resemblances upon which relationship is posited can differ or can be seen in different ways. For the reasons discussed in chapter 3 I regard the invention of the father as the foundation of kinship, that is, of a cultural order independent of, hence free to some extent from, the constraints of nature. Such freedom permits diversity and so enhances flexibility and adaptation.

Since the tie between the adult male and the offspring of the female cannot be known from biological evidence, it must be *made* known, and by analogy; hence it is the first form of relation that is the product of a concept. I maintain that once this metaphorical proposition has been made, the entire domain of relatedness that we now recognize as kinship becomes possible, and more than likely in a manner like that shown by Scheffler and Lounsbury in their semantic analyses of numerous human kinship systems from Pawnee to Siriono to Australian Aborigines. At the very least it seems that semantic and componential analyses of kinship systems are compatible with evolutionary theory (they permit a viable theory of origin and development), whereas alliance theory is not. We cannot propose a logically coherent theory for the evolution of alliance systems, since the logical origins proposed (the primacy of the group and the exchange made imperative by the incest taboo) are illogical.

The natural, determinate female/offspring tie among all mammals remains a universal model, or at least a source, for the metaphor of kinship. Insofar as this is so, it is not a cultural relationship, for it is not posited but self-evident and determined. It will always be women (rather than men) who will bear children, and individual children will always be related to individual women. In this sense, the female/infant tie is independent of culture and is transformed into a cultural relation of kinship by the application of the primary metaphor of fatherhood. The mother/child bond always has a double nature, being both natural (or determined) and cultural (hence undetermined) in human thought. This is not true of the male/infant tie, which is universally acknowledged culturally but not biologically; there are numerous examples of cultures in which the biology of paternity plays little or no part in the ideology of kinship and the definition of the father.

It may perhaps seem to be going rather far to argue that the female/infant bond is transformed by analogy with the father relation into the kinship relation "mother" in human society. Logically this must be so if a natural tie is to become a kinship, hence cultural, relationship. There is no evidence in any facet of human behavior that biologically determined activities directly produce culture—only that they provide the necessary conditions. And since the female/infant bond is invariable throughout the species, it would be hard if not impossible to explain how such invariance could give rise to the enormous variety of human kinship systems and practices that we see. Furthermore, the circumstantial evidence of comparisons with other primates (not to mention other mammals) indicates that human kinship must be founded on something more than biology, if not on an analogy with biology.

The "kinship" described for nonhuman primates is founded on the female/infant bond and tends to be relatively invariant in form for each species. Indeed, there appears to be little difference in kin group organization between species that have manifested such grouping. Equally, the incorporation of adult males as operative and instrumental members with structural positions defined by a bond analogous to that between females/infants and siblings is singularly lacking. But it is a hallmark of human kinship that both males (particularly males) and females have instrumental positions or status in all aspects—terminology, behavior, and groups—and it is equally noteworthy that all human kinship is bilateral. It is also characteristic of the species that position, status, and designation in kinship systems vary in a systematic way that cannot be shown to follow from universal factors that are biologically determined. Additional circumstantial evidence for the origin of kinship in the relation of father is the widespread subordination of the relation "mother" to that of "father" and, similarly, of the relation "sister" to that of "brother". If kinship reflected a biologically determined phenomenon, we would expect the reverse to be true, that females would ideologically as well as practically dominate the organization of relations. This is certainly true of nonhuman primates, since adult males remain structurally and operatively outside the kinship organization (though they play important emotional and supportive roles and are dominant in physical terms and relative to the total coresident population). But in human social organizations the adult male, whether father, grandfather, mother's brother, son, or sister's son, is operatively dominant. In certain descent systems a woman's child becomes the "property" of the male, the "father," and his relatives. In other systems a woman's child is more strongly and operatively related to her brother than to herself; the emotional feeling may of course be stronger in the female relationship, but the social and cultural emphasis is on the male relationship. And while such practice may make for contradiction in the actual course of social life between male and female and between male- and female-connected relatives, the fact that such conflicts do not overturn the system is in itself evidence for the dependence of kinship specifically, and perhaps more generally for culture, on the male. This male dominance, then, cannot have its roots in apparent physical superiority. Without the metaphor of the father, we are thrown back to the determined, biological connectedness of female and infant, and hence the invariant or restricted forms of organization and identification that we observe among nonhuman primates.

If we look outward from the father and from kinship systems, we can argue that with kinship is established the foundation for the positing of relatedness between human beings on the basis of metaphor. And this unitary domain of metaphor itself becomes a grounded, familiar realm of knowledge that may serve as a model and source for apprehending metaphorically other, nonhuman, phenomena. The metaphor of kinship assumes its polysemy. We relate, on the model of kinship, apparently disconnected natural phenomena, and at the same time we use apparent connections in nature as metaphors and models for our cultural conceptions, as when we speak of the family tree.

REFERENCES

Abercrombie, M., Hickman, C. J., and Johnson, M. L.
1957   *A dictionary of biology*. Harmondsworth, Middx.: Penguin.
Altmann, S. A., and Altmann, J.
1970   *Baboon ecology: African field research*. Chicago: University of Chicago Press.
Arendt, Hannah.
1978   *The life of the mind: Vol. 1. Thinking*. London: Secker and Warburg.
Austin, J. L.
1976   *How to do things with words*. (2nd ed.) London: Oxford University Press.
Bowlby, John.
1973   *Attachment and loss: Vol. 2. Separation—Anxiety and anger*. London: Hogarth Press.
Chance, M., and Jolly, C.
1970   *Social groups of monkeys, apes and men*. London: Thames and Hudson.
Chivers, D. J.
1977   The feeding behaviour of siamang. In *Primate ecology*, ed. T. H. Clutton-Brock. London: Academic Press.
Clark, J. D.
1976   The African origins of man the toolmaker. In *Human origins: Louis Leakey and the East African evidence*, ed. Glyn Isaac and E. R. McCown. Menlo Park, Calif.: W. A. Benjamin.
Darwin, Charles.
1899   *The origin of species by means of natural selection*. London: John Murray. (Originally published, 1859.)

171

1899 *The descent of man.* London: John Murray. (Originally published, 1871.)

DeVore, Irven, and Hall, K.R.L.
1965 Baboon ecology. In *Primate behavior: Field studies of monkeys and apes,* ed. Irven DeVore. New York: Holt, Rinehart and Winston.

DeVore, Irven, and Washburn, S. L.
1966 Baboon ecology and human evolution. In *African ecology and human evolution,* ed. F. Clark Howell and F. Bourlière. Chicago: Aldine.

Dunbar, R.I.M.
1977 Feeding ecology of gelada baboons: A preliminary report. In *Primate ecology,* ed. T. H. Clutton-Brock. London: Academic Press.

Engels, Friedrich.
1964 Ludwig Feuerbach and the end of classical German philosophy. In *Marx and Engels on religion,* ed. Reinhold Niebuhr. New York: Schocken Books. (Originally published, 1866.)

Errington, Karl F.
1974 *Karavar: Masks and power in a Melanesian ritual.* Ithaca, N.Y.: Cornell University Press.

Fossey, Diane, and Harcourt, A. H.
1977 Feeding ecology of free ranging mountain gorillas. In *Primate ecology,* ed. T. H. Clutton-Brock. London: Academic Press.

Fox, Robin.
1972 Alliance and constraint: Sexual selection in the evolution of human kinship systems. In *Sexual selection and the descent of man 1871–1971,* ed. Bernard Campbell. Chicago: Aldine-Atherton.

1975 Primate kin and human kinship. In *Biosocial anthropology,* ed. Robin Fox. London: Malaby Press.

Fromm, Erich.
1974 *The anatomy of human destructiveness.* London: Jonathan Cape.

Gallup, Gordon G., Jr.
1979 Self-awareness in primates. *American Scientist* 67.4:417–21.

Gardner, R. A., and Gardner, B. T.
1969 Teaching sign language to a chimpanzee. *Science* 165:384, 664–72.

Gibson, J. J.
1966   *The senses considered as perceptual systems.* Boston: Houghton Mifflin.

Godelier, Maurice.
n.p.   "Un problème qui n'est pas 'académique'"

Goldschmidt, Richard B.
1940   *The material basis of evolution.* New Haven, Conn.: Yale University Press.

Gould, Stephen Jay.
1977   *Ontogeny and phylogeny.* Cambridge, Mass., and London: Belknap Press of Harvard University Press.

Gregor, Thomas.
1977   *Mehinacu: The drama of daily life in a Brazilian Indian village.* Chicago: University of Chicago Press.

Gunn, Michael J.
1979   *Etak and other concepts underlying Carolinian navigation.* M.A. thesis, University of Otago, Dunedin, New Zealand.

Hobbes, Thomas.
1839–45   The English works of Thomas Hobbes of Malmesbury, ed. Sir William Molesworth: Vol. 3. *Leviathan.* London: John Bohn. (Originally published, 1651.)

Holloway, R. L.
1975   Early hominid endocasts: Volumes, morphology and significance for hominid evolution. In *Primate functional morphology and evolution,* ed. R. H. Tuttle. The Hague: Mouton.

Hume, David.
1888   *A treatise of human nature,* ed. L. A. Selby-Bigge. Oxford: Clarendon Press. (Originally published, 1739.)

Isaac, Glynn.
1971   The diet of early man: Aspects of archaeological evidence from lower and middle Pleistocene sites in Africa. *World Archaeology* 12:278–99.

Johanson, Don.
1979   A new history for man. *New Scientist* 81.1140:319.

Jolly, Alison.
1972   *The evolution of primate behavior.* New York: Macmillan.

Jolly, Clifford.
1970   The seed eaters: A new model of hominid differentiation based on a baboon analogy. *Man* (n.s.) 5:5–26.

Kant, Immanuel.
1970   Idea for a universal history with a cosmopolitan purpose.
            In *Kant's political writings,* ed. Hans Reiss. Cambridge:
            Cambridge University Press. (Originally published,
            1784.)
1974   *Anthropology from a pragmatic point of view.* Trans. M. J.
            Gregor. The Hague: Martinus Nijhoff. (Originally pub-
            lished, 1800.)
Katz, David.
1953   *Animals and men.* Harmondsworth, Middx.: Penguin.
Kummer, Hans.
1967   Tripartite relations in hamadryas baboons. In *Social com-
            munications among primates,* ed. Stuart Altmann. Chicago:
            University of Chicago Press.
1968   *Social organization of hamadryas baboons.* Chicago: Univer-
            sity of Chicago Press.
1971   *Primate societies: Group techniques of ecological adaptation.*
            Chicago: Aldine.
Lacan, Jacques.
1968   The mirror phase. *New Left Review* 51:71–77.
Lawick-Goodall, Jane van.
1974   *In the shadow of man.* London: Collins.
Leakey, Richard, and Isaac, Glynn.
1976   East Rudolf: An introduction to the abundance of new
            evidence. In *Human origins: Louis Leakey and the East Afri-
            can evidence,* ed. Glynn Isaac and E. R. McCown. Menlo
            Park, Calif.: W. A. Benjamin.
Lévi-Strauss, Claude.
1967   *Les structures élémentaires de la parenté.* Paris: Mouton.
Locke, John.
1824   A letter concerning toleration. In John Locke, Collected
            works, vol. 5. London: C. and J. Rivington. (Originally
            published, 1689.)
1960   *Two treatises of government.* Ed. Peter Laslett. Cambridge:
            Cambridge University Press. (Originally published,
            1690.)
Lorenz, Konrad.
1977   *Behind the mirror: A search for a natural history of human
            knowledge.* London: Methuen.
MacKinnon, J.
1974   The behaviour and ecology of wild orangutans (*Pongo
            pygmaeus*). *Animal behaviour* 22:3–74.

Monod, Jacques.
1972   *Chance and necessity: An essay on the natural philosophy of modern biology.* London: Collins.
Needham, Rodney.
1962   *Structure and sentiment: A test case in social anthropology.* Chicago: University of Chicago Press.
1964   Descent, category and alliance in Siriono society. *Southwestern Journal of Anthropology* 20:229–40.
1971   Remarks on the analysis of kinship and marriage. In *Rethinking kinship and marriage,* ed. Rodney Needham London: Tavistock.
1979   *Symbolic classification.* Santa Monica, Calif.: Goodyear.
Patterson, Colin.
1978   *Evolution.* St. Lucia, Queensland: University of Queensland Press, and London: British Museum (Natural History).
Pilbeam, David.
1972   *The ascent of man: An introduction to human evolution.* New York: Macmillan.
Popper, Karl R.
1972   *Objective knowledge: An evolutionary approach.* Oxford: The Clarendon Press.
Premack, A. J., and Premack, D.
1972   Teaching language to an ape. *Scientific American* 227:92–99.
Quine, Willard Van Orman.
1960   *Word and object.* Cambridge, Mass.: The M.I.T. Press.
Radcliffe-Brown, A. R., and Forde, Daryll.
1950   *African systems of kinship and marriage.* London: International African Institute.
Reynolds, Vernon, and Reynolds, F.
1965   Chimpanzees of the Budongo forest. In *Primate behavior: Field studies of monkeys and apes,* ed. Irven DeVore. New York: Holt, Rinehart and Winston.
Reynolds, Vernon.
1968   Kinship and the family in monkeys, apes and man. *Man* (n.s.) 3:209–23.
1975   How wild are the Gombe chimpanzees? *Man* (n.s.) 10:123–25.
Rousseau, Jean-Jacques.
1973   *A discourse on the origin of inequality.* Trans. and ed.

G.D.H. Cole. London: J. M. Dent. (Originally published, 1754.)

1973   *The general society of the human race.* Trans. and ed. G.D.H. Cole. London: J. M. Dent (Omitted from published version of *The social contract.*)

1973   *The social contract.* Trans. and ed. G.D.H. Cole. London: J. M. Dent. (Originally published, 1762.)

Sahlins, Marshall.

1977   *The use and abuse of biology.* London: Tavistock.

Schaller, George.

1965   Behavioral comparisons of the apes. In *Primate Behavior: Field studies of monkeys and apes,* ed. Irven DeVore. New York: Holt, Rinehart and Winston.

Schaller, George, and Emlen, J. T.

1966   Observations on the ecology and social behavior of the mountain gorilla. In *African ecology and human evolution,* ed. F. Clark Howell and F. Bourlière. Chicago: Aldine.

Scheffler, H. W., and Lounsbury, Floyd.

1971   *A study in structural semantics: The Siriono kinship system.* Englewood Cliffs, N.J.: Prentice-Hall.

Sebeok, Thomas A., and Umiker-Sebeok, Jean.

1979   Performing animals: Secrets of the trade. *Psychology Today* 13.6:78–91.

Simmel, Georg.

1950   *The sociology of Georg Simmel.* Ed. Kurt Wolff. New York: Free Press of Glencoe.

Straus, William L., Jr.

1949   The riddle of man's ancestry. *Quarterly Review of Biology* 24:200–23.

Szalay, F. S.

1975   Hunting-scavenging protohominids: A model for human origins. *Man* (n.s.) 10:420–29.

Teleki, Geza.

1973   The omnivorous chimpanzee. *Scientific American* 228:32–42.

Terrace, H. S.

1979   How Nim Chimpsky changed my mind. *Psychology Today* 13.6:65–77.

Tobias, Philip V.

1973   New African evidence on hominid phylogeny. Paper presented at the ninth congress of the International Union for Quaternary Research, Christchurch, New Zealand.

1975   Brain evolution in the Hominoidea. In *Primate functional morphology and evolution,* ed. R. H. Tuttle. The Hague: Mouton.

Turner, Victor W.
1967   *The forest of symbols.* Ithaca, N.Y.: Cornell University Press.

Wallace, V. A.
1975   Dietary adaptations of Australopithecus and early Homo. In *Primate functional morphology and evolution,* ed. R. H. Tuttle. The Hague: Mouton.

Washburn, Sherwood L., and Moore, Ruth.
1974   *Ape into man: A study of human evolution.* Boston: Little, Brown.

Wilson, E. O.
1975   *Sociobiology: The new synthesis.* Cambridge, Mass.: The Belknap Press of Harvard University Press.
1978   *On human nature.* Cambridge, Mass., and London: Harvard University Press.

Wilson, Peter J.
1967   Tsimihety kinship and descent. *Africa* 37.2:133–53.
1973   *Crab antics: The social anthropology of English-speaking Negro societies of the Caribbean.* New Haven, Conn.: Yale University Press.
1975   The promising primate. *Man* (n.s.) 10:5–20.
1978   La pensée alimentaire: The evolutionary context of rational objective thought. *Man* (n.s.) 12:320–35.

Wittgenstein, Ludwig.
1963   *Philosophical investigations.* (2nd ed.) Oxford: Basil Blackwell.

Wrangham, R. W.
1974   Artificial feeding of chimpanzees and baboons in their natural habitat. *Animal behaviour* 22:83–93.
1977   Feeding behaviour of chimpanzees at Gombe National Park, Tanzania. In *Primate ecology,* ed. T. H. Clutton-Brock. London: Academic Press.

Zuckerman, Solly.
1933   *Functional affinities of man, monkeys and apes.* New York: Harcourt, Brace.

INDEX

Acheulean culture: widespread, 31; and *H. erectus,* 31; and need for instruction, 32

Asymmetry of male/female relationship, 71–72; shown in idea of legitimacy, 71; shown in concept of adultery, 71; shown in African bridewealth, 72; in marriage, 79; and conflict of pair/primary bonds, 92

Attachment: of infant mother, 47; becomes two-way, 47. *See also* Primary bond

Australopithecus, 25, 26; species of, 27; in comparison with *Homo,* 28; cranial capacity of, 28

Baboons: relations between male and female, 49; social organization of, 50; diet, 116–17; and dietary barriers, 117; in competition with chimpanzees, 120, 122; as offering model for hominid diet, 129

Bananas: introduction of in Gombe, 120; as focus of competition, 120–21

Biology: limited in explanatory power, 2–3; related to context, 3

Birth, premature in humans, 46

Brain: and mind, 14, 112; as distinctive in human evolution, 15; and generalization, 16, 19, 134; and upright posture, 18, 22, 23; as executive coordinator, 19, 20, 23; and environmental problems, 24; growth rate of in humans, 36; increase in size of in *Homo,* 45, 134; extrauterine growth of in humans, 46; development of and reasoning, 111; of *Australopithecus,* 134; growth of as evolutionary response, 135

Caribbean society: and parentage, 72; ritualization of pair bond delayed in, 79–80

Caroline Islands: navigational instruction in, 145–46

Child: as pivot of relations, 95; as divided self, 95

Chimpanzees: as creatures less generalized than humans, 25; potential of, 26; and tools, 30; embryo and infant of similar to human, 34; and paedomorphism, 34–35; as sibling species, 35; age of sexual maturity in, 36; gestation of, 36; young of solve puzzles, 37; and life in groups, 40; lack of paternalism among, 48; social organization of, 51; females as core of group, 62, 63; and primary bond, 63; male of outside reproductive organization, 64; individuality of, 83, 84; absence of reciprocity among, 86; diet of, 117–18; eclecticism of, 118; and meat eating, 119; in conflict with baboons, 120, 122; and

Chimpanzees (*continued*)
bananas, 121–23; and problems with
humans, 122; taught by humans,
126–27, 143; able to learn but not to
teach, 143
Choice: as mechanism present in
nonhumans, 120; developed in humans, 120; and rationality, 120, 125;
and dietary preferences, 120, 135,
140; and competition, 135; and dietary competence, 140; and intention,
140; related to taste, 140–41
Clark, J. D., 32
Climate changes: in Miocene/Pliocene,
22–23; and ecological changes, 23; in
Pleistocene, 25
Competition: between baboons and
chimpanzees, 120, 122, 125; between chimpanzees and humans,
121–22, 125; and aggression, 125;
and predation, 125; between *Homo*
and *Australopithecus,* 133; between
hominids and pongids, 133; and food
choice, 135; and human intelligence,
142
Componential analysis, compatible
with evolutionary theory, 167
Conflict of interest: in pair and primary
bonds, 92–93; implicit in kinship, 98
Contradiction: of war of all against all,
4; of social contract, 4; between
needs and passions, 6; of individual
ownership, 7; of property and violence, 7; between desires and means,
8; society as, 9; of unsocial sociability, 9; as explanation of society, 9;
ahistorical, 9; of human evolution,
152; faced by individuals, 153; as the
human condition, 153
Cross-modality between senses, 140
Culture: emergence of in time, 1; associated with *Homo,* 1; genetically
immanent, 2; composition of,
11–13; extrasomatic, 30; begins with
H. habilis and *H. erectus,* 32; and
theories of cultural evolution,
154–55

Darwin, Charles: and theory of evolution, 2, 3, 10, 148; and neo-
Darwinian theory, 3; and specialization versus generalization, 37–38; on
African origins, 137; on human
evolution, 149
Deference of female to male, 93
Dentition: evolution of hominid,
131–32; functions of in primates,
131–32; relation of to tools and
weapons, 132; in relation to diet,
132
Dependency: of infant on mother, 47;
leading to attachment, 47
Diet: of nonhuman primates, 115–19;
and floral barrier, 118–19;
generalized in primates, 119; and
choice, 120; of *Australopithecus,*
128–29
Dimorphism of early hominids, 138
Division of labor and hunting, 139
Domination: of males by females, 65;
of culture by males, 65, 66–67, 139,
169; by males based on vulnerability,
67; of father/brother, 168
Duality of kinship relations, 94, 98
Durkheim, E.: and primacy of group,
157; on classification, 159

Ego in animals, 84
Environment: changes in considered as
initiative, 3; comprises species, 3;
considered a problematic, 3, 12; offers conditions for multiple solutions, 13; imposes limits, 13; exists
only in particulars, 18; change in
leads to problems, 24, 141; change in
and primates, 113; human alteration
of, 113, 114, 120, 141; and food,
114; and benign change, 125
Estrus: in nonhuman primates, 55; loss
of in humans, 55; and periodic activity, 88
Evolution: occurs if potential becomes
actual, 2; as more than biology, 2–3;
of particular species considered a fiction, 3; as a seamless web, 3; and

interrelation of conditions of setting and organism, 3, 12, 150; continuities of, 12, 42, 148, 156; mosaic, 25; specifics of in humans, 42–43; theory of unique to humans, 148; as response to problems, 150; and purpose, 150–51; theories of in humans, 154–55

Father: as product of primary and pair bond, 59–60, 95; allows for transferral of relations, 60; as independent of biology, 60, 66; as origin of human kinship, 64, 65, 95, 166; definition of, 65; as fragile institution, 66; identity of guaranteed, 71; concept of perpetuated by incest taboo, 74; dependence of, 91–92; as node of kinship, 95; as divided self, 95

Female: pivotal position of, 94; sees herself as divisible, 94; duality of in bonds, 94; as first to share, 94; as first to promise, 94; as cultural as male, 96

Food: as reward, 114; as problem, 114; and rationality, 114; and problem solving, 126

Fox, R., ideas of, about kinship disputed, 53, 62, 63

Freud, S.: on incest taboo, 74, 78; on Oedipus complex, 77

Function. See Teleology

Future: concept of, 103; created by promise, 106, 145; and language, 109

Gathering: and hunting, 66; and male/female roles, 66, 96, 138; cooperation and sharing in, 96, 139

Generalization: described by Rousseau, 5; described by Hume, 8; indicated by human distribution, 15; characteristic of human primate, 15; defined as not specialized, 16; and cerebral development, 16, 18, 19; and posture/gait, 17–18; humans as most generalized creature, 18, 133;

weakness and strength of, 19; as more demanding of thought, 19; evidence of in H. habilis and H. erectus, 32–33; contradicts Darwinian theory, 37, 152; and neoteny, 37, 153; adaptive advantage of, 38; not confined to morphology, 39; of social organization in humans, 40; and uncertainty, 43; of primate social organization, 52; of sexual nature, 66–69, 77; of ritual, 108; of kinship, 108, 111; of H. erectus, 130, 131; evolution of in hominids, 133; of dietary competence, 133

Genes: and mutation considered as initiative, 3; and evolutionary theory, 155

Gibbons: family group of, 40; paternal care by, 48; social organization of, 49–50; diet of, 116

Goodall, Jane, 21

Gorillas: social organization of, 51; diet of, 115

Hanuman langurs: paternal care by, 48

Hegel, G. W. F., 131; and holism, 157

Helplessness: in human infant, 46; consequences of, 46–47

Hobbes, Thomas, 9; resolutive-compositive method of, xi; and contradiction, 4; and war of all against all, 4

Hobbesian war, 4, 66; ahistorical, 10; and sexuality, 89; and kinship, 99

Holism, 157–58

Homo: species of, 27; cranial capacity of, 29; and tools, 29–30; H. habilis, character of, 30; H. erectus widespread, 31; H. erectus and Acheulean culture, 31; H. erectus, cranial capacity, 31; infantilism of, 34, 35; sexual maturity of, 36; gestation of, 36; helplessness of at birth, 36; comparability of species of, 40; generalized social organization of, 40–41; and the invention of kinship, 41

Hume, David: experimental method of, xi; and argument for conjunction of infirmity and necessity, 8; on nature of property, 8; on instability of possession, 9
Hunting: and gathering, 66; and male/female roles, 66, 96, 138; cooperation and sharing in, 87, 96, 139; development of in chimpanzees, 125, 126; and aggression, 139

Identity: as showing fitness, 154; relativity of, 157; of individual distinct from group, 160
Incest taboo: prohibits sexual relations between kin, 73; as a public phenomenon, 73, 102; enables reproduction of kinship, 74; innateness of considered irrelevant, 75; on mother/son relations, 75; on brother/sister relations, 75–76; Lévi-Strauss's ideas about criticized, 75–76, 158–59; in matrilineal societies, 76; on father/daughter relations, 77; as concerning principles, 77–78; permits principle of kinship, 78, 102; focuses on primary bond, 79; maintains integrity of promise, 102
Individualism, 158
Individuality as self-consciousness, 85
Individuation: of animals, 83; defined, 84–85; and sexual variation, 88
Infant helpless and dependent in humans, 46
Intention: formulation of, 140; in relation to food, 140

Johanson, Don, 27

Kant, I., 4; on unsocial sociability, 9, 69
Kibbutz and incest taboo, 73–74
Kinship: invented by early hominids, 41; universal to humans, 41; diagnostic of human being, 42; from primary and pair bonds, 61; as generalized mode of organization,

61, 108, 112; adaptive advantages of, 70; and problem of perpetuation, 70–71; not guaranteed, 71; as prerequisite for incest taboo, 73, 75, 158, 159; reproduced by incest taboo, 74; guarded by incest taboo, 78; viable only while adaptive, 81–82; articulated by offspring, 95; and promise, 98; produced by technique, 104; and language, 108; flexibility of, 112, 168; and identity, 157; existence of doubted, 157–58, 162; terms considered polysemic, 162; as metaphor for relatedness, 162, 164; and concept of resemblance, 163, 164; as a mode of thought, 164; as a term is analogous, 165, 166

Laing, R. D., 85
Language: and animal thought, 85; and expression of consciousness, 85; in conditions of promise and taboo, 108; arises in subjective conditions, 108; in relation to tools, 108; and motor capacities, 108; and ritual, 108; as generalization, 108; theories of origin of, 109; and idea of future, 109
Lawick-Goodall, Jane van. See Goodall, Jane
Leakey, L. S. B. and M., 27
Locke, John, 4; on human freedom, 7; on individual ownership, 7; and labor theory of property, 7; and conflict of property and violence, 7

Maori identity, 154
Marmoset, paternal care by, 48
Marriage: defined as cultural, 79; created through ritual, 79; public nature of, 79; as complement of incest taboo, 80; as transitional rite, 80–81; perpetuates kinship, 81
Marx, Karl, xii; and human acts relative to material conditions, 10; and holism, 157

Matrilineages: always include males, 63; as cumulative primary bonds, 63
Meat eating: among chimpanzees, 118; evidence for in early hominids, 129–30; evolution of in hominids, 136; in relation to brain development, 136–37, 138; and escape from floral limits, 137; and male specialization, 138; and sharing, 139; and ritual, 139
Mehinacu and interest in sex, 90
Metaphor: as prime mode of human thought, 162; originates in kinship, 162, 169; of kinship in relation to games, 163
Migration as adaptation, 137
Mother: as biological relation, 165–66; transformed into kinship relation, 167

Ndembu and ritual, 104–05
Neanderthal and ritual, 105
Neoteny: and *H. sapiens,* 34, 35; as result of retarded growth, 35, 37; consonant with generalization, 37
Nuer and fatherhood, 72

Odd-job words, 158, 162
Oldowan culture: associated with *H. habilis,* 29, 137; simple tools in, 32
Ontogeny: appears in embryo, 34; of humans, marked by slow growth, 35; and phylogeny, 153
Orangutans: solitary, 40; social organization of, 50; diet of, 115

Paedomorphism: of adult to embryo within species, 34; of humans, 34–35
Pair bond: and sexual attraction between male and female, 58; and overlap with primary bond, 59, 97; as biologically based, 70; institutionalized by marriage, 79; as basis for attribution of fatherhood, 80; and sexuality, 91; in relation to primary bond in humans, 91, 95

Philosophical anthropology, 4; and relation to facts, 10
Phylogeny: appears in fossil record, 33–34; of humans tends to generalization, 44
Prescriptive marriage alliance: as basis of kinship, 157; incompatible with ontology, 167
Primary bond: between mother and infant, 47; becomes reciprocal, 47–48; lacking between male and infant, 48; as natural mammalian trait, 48, 168; foreshadowed in primates, 59; and overlap with pair bond, 59, 97; biologically based, 70; as focus of incest taboo, 79; in relation to pair bond in humans, 91, 95; as basis of nonhuman "kinship," 168
Primates: showing wide but restricted distribution, 16; not carnivorous, 16; include some specialized species, 16, 17; variation in organization of, 40; disinterest in mixed-sex groups of, 57
Problem solving: as measure of intelligence, 113; and procurement of food, 114; and environmental change, 120; and chimpanzees and bananas, 121; and humans and chimpanzees, 121–23; and maximization of self-interest, 123–24; evolution of, 126; and rationality, 127; and meat eating, 138; as basis of evolution, 150; and teleology, 152
Promise: need for in relation between bonds, 94; and idea of self in future, 94; as kinship, 97; as a tool, 98; in relation to separation, 100; in overcoming of separation, 100–01; as convention, 101; and trust, 101; joins individual with social, 103; mystery of, 106; as creator of future, 106, 145; and the symbol, 106–07, 108, 145; as condition of teaching, 145
Property: Locke's labor theory of, 7, 8; transferability of, 8; and instability of

Property (*continued*)
possession, 9; as self-possession, 93;
and children, 93

Rationality: in primates, 112; developed in humans, 112–13; and environmental change, 113, 141; and
food, 114, 127; and choice, 120;
chimpanzee and human, 126; and
meat eating, 136–37; analyzed,
140–41; and theorizing in chimps,
142, 143
Reduction: of human nature to irreducibles, xi; not viable subjectively, xii
Reference: of male to female and offspring, 93
Reproduction: and sexuality, 91; and
problem of organization, 111; and
kinship, 111
Ritual: as technique of social reproduction, 104–05; as mode of transformation, 104–06; as prerequisite for social life, 105; and symbol, 107; as
performative action, 107; not symbolic, 108; and language, 108; and
sharing of meat, 139
Rousseau, J. J., 4; on amorality of
natural man, 4; on eclecticism of
human nature, 5; on overcoming obstacles, 6; on contradiction between
needs and passions, 6; and influence
on Kant, 9
Rules instituted to control sex, 57

Sarah, chimpanzee using symbols, 141
Self-consciousness: as ownership, 85;
expressed as division, 85, 98, 153;
and language, 86; and sharing, 86;
and idea of future, 86–87; and sexuality, 89, 90; and pair and primary
bonds, 91; by reference and deference, 93; through promise by
female, 94; in female, 95; in male,
95; in child 95, 153; of humans,
species-specific, 96; and kinship, 98
Separation: of relationship, 99; not
guaranteed, 99, 152; fear of, 99–

100; obligation of, 100; on condition
of renewal, 100; and understanding,
100; and promise, 100–01; of means
from ends, 152
Sex: specialization according to in primates, 49; specialization of in primates, 55; generalized in humans,
56, 88; and basis for pair bond, 58,
91
Sexual activity: defines male/female relationship among humans, 56, 58;
human capacity for continuous, 56,
68, 88; interest in considered specifically human, 67–68, 89–90; and
techniques learned, 68; subject to
rules, 68; as focus of human reflex-
iveness, 69; periodic in nonhumans, 88; and self-consciousness,
90; and dyadic quality of sexual relationships, 90–91
Sexuality: lacking in nonhumans, 87; as
defining feature of humans, 88; variations of, 88; as source of human
individuation, 88; as basis of attraction, 89; as basis of individual selection, 89; related to selfishness, 90;
and reproduction, 91
Sharing: and self-consciousness, 86;
and reciprocity, 86; and obligation,
86; absent in nonhuman primates,
86, 125; implies future, 86–87; and
division of labor, 87; theories of inadequate, 87; and sexual relations,
90; as reference and deference, 93;
of offspring by female, 93, 94; and
promise, 94; by individual of self, 99;
and hunting, 139; and ritual, 139
Siriono test case in kinship studies, 161
Social contract, 4
Social organization: as an artifact,
40–41; generalized in humans, 41,
52, 61–62; not genetically determined in humans, 52; significance of
among primates, 53–54; specialized
among nonhumans, 62; and origin of
symbol, 159
Sociobiology: theoretical claims of, 2;

on genes as bearers of human potential, 2; maintains that culture is immanent, 3, 155; and evolutionary theory, 155

Specialization: of primates, 16–17; imposes limitations, 17; and human posture, 17; and Darwinian theory, 37; of human brain, apparent, 134; and human evolution, 149

Symbol: and promise and taboo, 106; and relation to future, 107; necessity of, 107; as technique of ritual, 107; as language, 108; and teaching, 144; originates in social group, 159

Taboo: as sanction of promise, 101; rigidity of, 101; presupposes objects, 102; not a definition, 102; functions of, 102; maintains integrity of self, 102; joins individual with social, 103; creates that which is not, 106; and symbol, 106–07, 108

Taxonomy: of early hominids, 128; evolutionary theories and, 154

Teaching: of chimps by humans, 126–27, 141, 142; as means of human adaptation, 143–44; subjective, 144; as indirect or symbolic, 144, 145, 146; founded on promise, 145; as means of cultural reproduction, 146

Teleology: typical of hominids, 20, 51; defined within evolution, 150; of evolution, 150–51; implicit in study of evolution, 151

Thought: necessary for material and social artifacts, 14; demanded by generalization, 19; influence of uprightness on, 26; as metaphor, 162; as kinship, 164

Tools: first evidence of, 29; reflexivity of, 30–31

Transformation: of individuals, 104; in ritual contexts, 104, 105

Transitivity: of relationships, 60, 64; as means to kinship, 60–61; among primates, 64–65

Trobriand Islands, 21

Tsimihety, trial marriage among, 72, 80

Upright posture: as generalization, 17; as apparent specialization, 20; and connection with encephalization, 21, 23; consequences of, 21–22; problems of, 25; contributes to thought, 26

Washoe, chimpanzee learning ASL, 141

Wilson, D. H., brother author, 185

Wilson, E. O., 2

"Origin of man now proved.—Metaphysics must flourish.—He who understand baboon would do more toward metaphysics than Locke"—from Darwin's Notebooks on Man, Mind and Materialism (as quoted in Howard E. Gruber, *Darwin on Man: A Psychological Study of Scientific Creativity*, New York: E. P. Dutton, 1974).